HIGHER

GEOGRAPHY
GLOBAL ISSUES

SECOND EDITION

Calum Campbell
Ian Geddes

HODDER
GIBSON
AN HACHETTE UK COMPANY

The Publishers would like to thank the following for permission to reproduce copyright material:

Photo credits
Image used on Chapter opener **pages 1**, **35** and **88** and in Task boxes throughout © Romolo Tavani – Fotolia; **p.3** (top left) © by paul – Fotolia, (top right) © ChiccoDodiFC – Fotolia; **p.4** © mady70 – Fotolia; **p.9** © eunikas – Fotolia; **p.10** © Bureau of Reclamation; **p.11** © John Miles/ The Image Bank/Getty Images; **p.18** © ChinaFotoPress via Getty Images; **p.19** © Richard Jones/Shutterstock; **p.21** © Wang Jiaman/Chine Nouvelle/Sipa/Shutterstock; **p.27** © Bureau of Reclamation; **p.28** (top left) © STR/AFP/Getty Images, (bottom left) © Julie Dermansky/ Corbis via Getty Images; **p.29** © Bureau of Reclamation; **p.33** © BANARAS KHAN/AFP/Getty Images; **p.36** © Noah Addis/Corbis; **p.40** (top left) © Copyright Leslie Barrie/http://www.geograph.org.uk/photo/3663271/http://creativecommons.org/licenses/by-sa/2.0/, (bottom left) © by Mohamed Hossam/Anadolu Agency/Getty Images; **p.52** © Shutterstock / Vlad Karavaev; **p.63** © Hemis / Alamy Stock Photo; **p.58** © Str/EPA/Shutterstock; **p.60** (top) © ruticar – Fotolia, (bottom) © Raimond Klavins/stock.adobe.com; **p.62** © Henrik Larsson – Fotolia; **p.64** © Spencer Platt/Getty Images; **p.67** © Eye Ubiquitous/Shutterstock; **p.76** © Asia File / Alamy; **p.77** © Mile 91/Ben Langdon / Alamy Stock Photo; **p.79** © World Health Organization/P S. Hollyman, 2014; **p.81** © VCG via Getty Images; **p.88** © frog-travel – Fotolia; **p.89** (left to right) © Goinyk Volodymyr – Fotolia, © Les Cunliffe - Fotolia.com, © Digital Vision/Getty Images; **p.91** © Getty Images/iStockphoto/Thinkstock; **p.96** © Digital Vision/Getty Images; **p.98** © James Thew – Fotolia; **p.99** (top left) © StockTrek/Photodisc/Getty Images, (bottom left) © ricknoll – Fotolia; **p.102** © Shutterstock / rweisswald; **p.105** © U.S. Geological Survey/photo by K Jackson; **p.109** © filtv – Fotolia; **p.116** © Getty Images/Moment Open; **p.120** (top) © The Asahi Shimbun via Getty Images, (bottom) © Nicolo E. DiGirolamo, SSAI/NASA GSFC, and Jesse Allen, NASA Earth Observatory; **p.129** © Torsten Blackwood/AFP/Getty Images; **p.131** (top left) © Patrick Aventurier/Gamma-Rapho via Getty Images, (top right) © ton koene / Alamy Stock Photo, (centre right) © Torsten Blackwood/AFP/Getty Images; **p.140** (top to bottom) © Fotimmz – Fotolia, © kenxro - Fotolia.com, © Michele Burgess / Alamy, © Jean-Christophe Verhaegen/AFP/Getty Images, **p.141** (top to bottom) © nofear4232 – Fotolia, © Doug Houghton / Alamy, © De Agostini/Getty Images, © B Lawrence / Alamy.

Every effort has been made to trace all copyright holders, but if any have been inadvertently overlooked the Publishers will be pleased to make the necessary arrangements at the first opportunity.

Although every effort has been made to ensure that website addresses are correct at time of going to press, Hodder Gibson cannot be held responsible for the content of any website mentioned in this book. It is sometimes possible to find a relocated web page by typing in the address of the home page for a website in the URL window of your browser.

Hachette UK's policy is to use papers that are natural, renewable and recyclable products and made from wood grown in well-managed forests and other controlled sources. The logging and manufacturing processes are expected to conform to the environmental regulations of the country of origin.

Orders: please contact Hachette UK Distribution, Hely Hutchinson Centre, Milton Road, Didcot, Oxfordshire, OX11 7HH. Telephone: +44 (0)1235 827827. Email: education@hachette.co.uk Lines are open from 9 a.m. to 5 p.m., Monday to Friday. You can also order through our website: www.hoddereducation.co.uk. If you have queries or questions that aren't about an order, you can contact us at hoddergibson@hodder.co.uk

First published in 2015 © Calum Campbell and Ian Geddes
This second edition published in 2019 by
Hodder Gibson, an imprint of Hodder Education,
An Hachette UK Company
50 Frederick Street
Edinburgh, EH2 1EX

Impression number	5	4	3	2
Year	2023	2022		

Cover photo © Anton Balazh/Shutterstock.com
Illustrations by Integra Software Services Pvt. Ltd., Pondicherry, India
Typeset in DIN 10/13 by Integra Software Services Pvt. Ltd., Pondicherry, India
Printed in the UK

A catalogue record for this title is available from the British Library

ISBN: 978 1 5104 5775 1

SCOTLAND EXCEL

We are an approved supplier on the Scotland Excel framework.

Find us on your school's procurement system as *Hachette UK Distribution Ltd* or *Hodder & Stoughton Limited t/a Hodder Education*.

MIX
Paper | Supporting responsible forestry
FSC™ C104740
www.fsc.org

Contents

Acknowledgements

I'd like to thank all those people who have encouraged and supported me through the writing of this book and in no particular order of importance: my wife Wendy and daughter Laurin; Nancy and George Davidson; Ian Geddes, my co-author; Jim Knox, Alan Johnstone, Bob Reid, Christine Campbell, Brady Robertson, Laura Kerr and Martyn Olesen; all former 'Byristas', with huge appreciation to the kindness and understanding of Eddie and the late Evelyn Carden; all my friends and their families; never forgetting Jim, Muriel, Charlie and Jane.

Calum Campbell

Whilst writing this book has been a pleasure, I am delighted to acknowledge the support and encouragement from my wife, Susan, and daughters Amelia and Susan. Thanks too to my co-author Calum for his encouragement to discuss matters over all-day breakfasts or tea and scones. Over the years I have been privileged to teach many students of geography and it is for them that this book has been written. Thanks also to Katie, Lucy, Blair and Donny, the future generation of geographers! Finally, the support and advice from the staff at Hodder Gibson in Glasgow has been immense. Thank you all.

Ian Geddes

Introduction to Higher Geography Global Issues

About this book

This book is one of two core text books and has been written to help you prepare for the SQA Higher Grade Geography Examination. The focus of this core book is Global Issues and it contains three of the key topics:

- River Basin Management
- Development and Health
- Global Climate Change

The three topics have been selected to reflect those requested by most schools. A fourth topic, 'Energy', is not covered in the content of this book. From our research very few schools have stated that they will include Energy.

The three global issues in this book have been broken down into a number of smaller descriptive topic areas:

River Basin Management

- Physical characteristics of a selected river basin
- Need for water management
- Selection and development of sites
- Consequences of water control projects

Development and Health

- Validity of development indicators
- Differences in levels of development between developing countries
- A water-related disease: causes, impact, management
- Primary healthcare strategies

Global Climate Change

- Physical and human causes
- Local and global effects
- Management strategies and their limitations

Aims of the Higher Geography course

The Higher Geography course develops learners' understanding of our changing world and its human and physical processes in local, national, international and global study contexts.

In our fast-changing and increasingly complex world, an underlying theme throughout this book is to encourage learners to interact with their environment.

This book will enable you to develop:

- a wider range of geographical skills and techniques
- an understanding of the complexity of ways in which people and the environment interact in response to physical and human processes at the local, national, international and global scales
- an understanding of spatial relationships and of the complexity of the changing world in a balanced, critical and sympathetic way
- a geographical perspective on environmental and social issues and their significance
- an interest in, understanding of, and concern for the environment and sustainable development.

We hope that you agree with the significance and importance of these aims.

Structure of Higher Geography Global Issues

Each chapter in this book shares a broadly similar format and contains the following features:

- Subject content
- Background content considered to be relevant to full understanding of each section
- Key case studies
- Tasks to test knowledge, understanding and application of skills
- Opportunities for reflection
- Opportunities for further research.

A glossary of key terms is also provided at the end of the book.

Examination structure

The source of the information contained in this section is the Scottish Qualifications Authority – www.sqa.org.uk. You are recommended to obtain a copy of the latest course specification and to refer to the website for updates, old question papers and information.

The SQA Higher Geography course contains three mandatory sections:

- Geography: Physical Environments
- Geography: Human Environments
- Geography: Global Issues.

However, within this you do have some scope for choice. As stated earlier, there are two books in this series. The other book, *Physical and Human Environments*, contains the following key topics:

Physical Environments

- Atmosphere
- Hydrosphere
- Lithosphere
- Biosphere

Human Environments

- Population
- Rural (land degradation and management)
- Urban (change and management)

The context and examples in *Physical and Human Environments* are from within urban and rural areas in developed and developing countries. The way it will be taught in your school or college will be through general background skills and information combined with case studies for your chosen areas of study.

How is Higher Geography assessed?

The course is assessed by three components:

- Component 1: Question paper 1, Physical and human environments (100 marks*)
- Component 2: Question paper 2, Global issues and geographical skills (60 marks*)
- Component 3: Assignment (30 marks*)

Total marks available 190*

* SQA has chosen to *scale* the question paper marks. This means that although Question paper 1 is marked out of **100** and Question paper 2 is marked out of **60**, there is a statistical process that will reduce the overall value of each paper to **50** and **30** marks, respectively.

The course assessment is graded **A–D**. Your grade is determined on the basis of your total mark for the three course components added together.

Components 1 and 2: Question papers (160 marks)

The examination is set and marked by SQA and appointed specialist teachers of geography. There are two question papers. You are required to answer in an 'extended-response' manner (i.e. you must write in sentences, not lists or bullet points) using knowledge, understanding and skills you have acquired during the course.

This book covers in detail the content for Question paper 2.

Question paper 2: Global issues and geographical skills (60 marks)

This question paper has two sections:

- Section 1 Global issues is worth 40 marks and consists of extended-response questions. You should choose two from the four questions. Each question is worth 20 marks.
- Section 2 Application of geographical skills is worth 20 marks and consists of a mandatory extended-response question. You should apply geographical skills acquired during the course. The skills assessed in this section include the use of mapping, numerical and graphical information.

You have **1 hour and 10 minutes** to complete this question paper.

Your teachers will guide you regarding standards, structure, requirements and marking. The SQA publishes 'Specimen question papers' including marking instructions at www.sqa.org.uk.

Question paper 1: Physical and human environments (100 marks)

This question paper has two sections:

- Section 1 Physical environments
- Section 2 Human environments

Each section is worth 50 marks and consists of extended-response questions. You should answer **all** the questions in each section.

You have 1 hour and 50 minutes to complete this question paper.

The content for Question paper 1 is covered in the accompanying Hodder Gibson book, *Physical and Human Environments* (ISBN: 9781510457768).

Remember that for your final grade SQA will *scale* the question paper marks down to 80 marks.

Component 3: Assignment (30 marks)

Geography is not only about reading from textbooks. It is also about researching, thinking for yourself and applying all that you have covered in the classroom into a 'practical' setting. The assignment is your opportunity to demonstrate the application of these skills, knowledge and understanding within the context of a geographical topic or issue. The time allocated for writing up the assignment, under exam conditions, is 1 hour and 30 minutes.

Although this book includes ideas for assignments, the Hodder Gibson book, *How to Pass National 5 and Higher Assignments: Geography* (ISBN: 9781471883088) by Susie Clarke covers the skills in considerable detail.

The assignment allows you to show your skills, knowledge and understanding by:

- identifying a geographical topic or issue
- carrying out research, which should include fieldwork where appropriate
- demonstrating knowledge of the suitability of the methods and/or reliability of the sources used
- processing and using a range of information gathered
- drawing on detailed knowledge and understanding of the topic or issue
- analysing information from a range of sources
- reaching a conclusion supported by a range of evidence on a geographical topic or issue
- communicating information.

Examination techniques

The Higher examination set by SQA is fair and is based on a clear specification. SQA has a long history of delivering quality examination standards. A Higher Geography pass is recognised as a useful and worthwhile qualification, so it is worth time and effort

to develop your skills and knowledge to ensure you achieve a pass.

Figure 0.1 shows the examination jigsaw – recognise the six parts required to put together so that you pass.

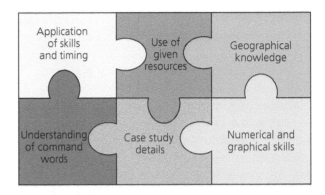

▲ **Figure 0.1** Examination jigsaw

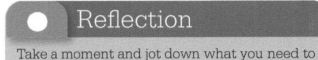

Reflection

Take a moment and jot down what you need to do to **fail** an exam!

You probably felt it strange to consider what you need to do to fail but hopefully you have come up with some of these suggestions:

- failure to revise
- failure to manage your time
- failure to turn up for the exam on the right day or time
- failure to follow the task that you are set.

All of this is under your control!

Command words

In the Geography examination there are a number of possible **command words** (instruction words) that tell you how to answer a question. The course assessment notes say that you have an 'opportunity to demonstrate:

- using a wide range of geographical skills and techniques
- describing, explaining, evaluating and analysing complex geographical issues, using knowledge and understanding which is factual and theoretical, of the physical and human processes and interactions at work within geographical contexts on a local, regional and global scale.'

Be aware that there is some *overlap* between the key command words highlighted over the page.

1 **Describe:** identify distinctive features and give descriptive, factual detail. This is one of the most widely used command words. The marking instructions also allow you to further develop these points, providing more depth.

2 **Explain:** here you are asked to provide the causes of a feature, issue or pattern. You need to show an understanding of processes and sources. For example, 'Explain the conditions and processes involved in the formation of a corrie.'

3 **Evaluate:** weigh up several options or arguments and **come to a conclusion** with regard to their relative importance/success/impact. Your judgement is important. Provide an insight into the thinking that led to your decision or conclusion. For example, 'Evaluate the effectiveness of various methods used to control the spread of malaria.'

4 **Analyse:** you need to be able to break down the content into its constituent parts in order to provide an **in-depth account**. These questions involve relationships between and within topics, and recognise implications, links, variety of views and consequences. For example, 'Analyse the impact of migration on either the donor country or the receiving country.'

It is possible that a further *three* commands could be used in the question paper:

5 **Account for:** in this style of question you are asked to give reasons for trends, issues, decisions and alternative actions. Account for questions, like many geographical tasks, are often based on issues which can be tested or may be subject to different views.

6 **Discuss:** the key idea here is to explore various different ideas about a project, an impact or a change. Once again, you are expected to be able to review alternative scenarios. Although you can express your own views, you must also show an awareness of other, contrasting views.

For example, 'Discuss the possible impact of global warming throughout the world.'

7 **To what extent:** here you are asked to consider the impact of a plan, strategy or programme and to form a view on the success or failure of that programme. In geography, there is seldom 100 per cent agreement about an outcome. Your awareness of the competing interests and values is important. For example, 'To what extent has the Colorado River Management Project achieved its aims of flood control and power generation?'

Process model

Geographers have a way of going about their business. Geography is a social science or a social subject, and as such it has a logical and sequential way of looking at the world and the human and physical interactions. Figure 0.2 shows the process model that illustrates the way geographers approach issues.

Marker's perspective

What do markers look for? They look for relevance – if it is not asked for, then do not include it in your answer. Markers love detail, so work really hard at your case studies. Higher Geography is only marked in full marks, there are no half marks. You need to write in sentences. If you look at a set of 'markers' instructions' (go to www.sqa.org.uk) you will see how marks are awarded. For one mark you need to provide a developed (detailed) point, so get into the habit of writing sentences that show this.

Markers like to see that students have attempted diagrams to illustrate an answer. They will first mark all the writing and only then give additional marks for new information in and around a diagram. (Be aware though that if a question asks you specifically to draw a diagram you *must* do so as this forms part of the answer that will gain marks.)

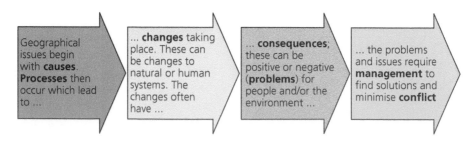

Geographical issues begin with **causes**. **Processes** then occur which lead to ...

... **changes** taking place. These can be changes to natural or human systems. The changes often have ...

... **consequences**; these can be positive or negative (**problems**) for people and/or the environment ...

... the problems and issues require **management** to find solutions and minimise **conflict**

▲ **Figure 0.2** The geographical way of doing things!

Markers are fair and want you to do well. They like an answer that has a clear structure and can be easily read. Do not repeat yourself and do get into the habit of using words such as, 'this is because ...' or 'an example of this is ...' Remember: you cannot lose marks, you only gain them.

To summarise, marks are awarded when your answers are:

- relevant to the issue in the question
- developed responses (by providing additional detail, reasons or evidence)
- used to respond to the command words/demands of the question (e.g. evaluate, explain, analyse).

Revision strategies

There are lots of good revision books and lots of websites and your teachers will give you advice. Here are our top ten revision tips.

1 Create your own personal study space.
2 Get organised: cards, pens, folders, highlighter pens, notes ...
3 Use relaxation techniques to get you into the mood for revision.
4 Actively read and write notes.
5 Use your notes and past papers.
6 Be creative in your revision, for example try using mind maps.
7 Leave your study space organised for the next session.
8 Take breaks and time out and reflect on your revision session.

9 Put in plenty of quality revision time.
10 Get support from family and friends.

▲ **Figure 0.3** 'I've passed!'

Examination techniques

- Be organised and arrive at the exam hall in plenty of time, with all the equipment needed.
- Make sure that you have eaten and have brought water to drink.
- Follow all the instructions and answer the correct number of questions from each section.
- Remember relaxation techniques.
- Remember that you have done the work so do not panic, read the question again and stay calm.
- Keep your answer legible.
- Match the length of each answer appropriate to the number of marks on offer.
- Be aware of time and manage it to maximum effect.
- At the end, check all your answers and add details.

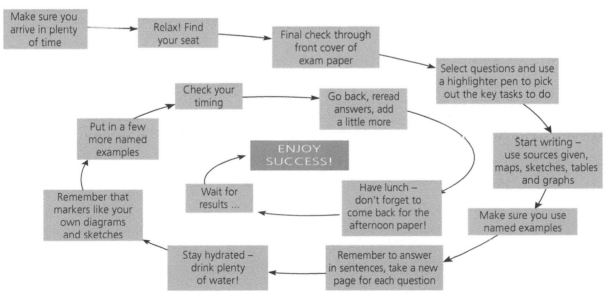

▲ **Figure 0.4** Tips to ensure success

Introduction to River Basin Management

The SQA course assessment specification lists the following topics to be covered in River Basin Management:

- Physical characteristics of a selected river basin
- Need for water management
- Selection and development of sites
- Consequences of water control projects.

Case study examples from both the developed and developing world will be introduced throughout this chapter.

There is an overlap in content, knowledge and skills in this section, 'River Basin Management', with the Physical Environment section 'Hydrosphere', namely the requirement to understand the hydrological cycle within a drainage basin. In view of that fact, we will begin by summarising some of the features of the hydrological cycle and some river features. We will therefore begin with the basics.

1.1 Drainage basin hydrological cycle

Key --- Main watershed --- Catchment area of a sample
  → Stream   large stream within the basin
★ Sources of streams, i.e. springs, bogs, lochs
○ Confluence point, i.e. where two or more streams meet
▭ Groundflow/throughflow of water

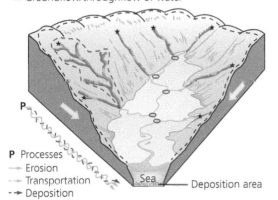

P Processes
→ Erosion
-→ Transportation
-→ Deposition

These river processes all take place along the length of rivers, i.e. from source to mouth

▲ **Figure 1.1** Main features of a drainage basin

All river basins are different yet share a number of similar physical characteristics. A drainage basin is the catchment area of a river and of all of the tributaries. It is the area from which a river system obtains its water. An imaginary line can be drawn around this area and this watershed delimits and defines one drainage basin from another.

Generally the watershed follows a ridge of high land (Figure 1.2). Any rain falling either side of this watershed will drain into different river basins.

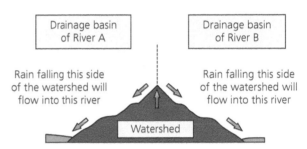

| Drainage basin of River A | Drainage basin of River B |

Rain falling this side of the watershed will flow into this river

Rain falling this side of the watershed will flow into this river

Watershed

▲ **Figure 1.2** Watershed

Of the water falling over land, 83 per cent will drain into the oceans. The other 17 per cent drains to internal endorheic basins with no sea outlet, such as the Great Basin in North America.

The drainage basin hydrological cycle is what is called an open system with inputs and outputs. 'Open' in the sense that the precipitation can enter the basin from outside, 'inputs' because rain and snow will fall into the system and 'outputs' because the rain and snow can leave the system.

The amount (volume) of water in our planet is fixed. It does not enter into or leave planet Earth. This is recognised as a system (Figure 1.3). What does change is the property of that water, whether it is a gas, a liquid or a solid. The simple concept behind the cycle is that water transfers from the oceans into the atmosphere over the land, finally returning to the oceans. The processes involved are basic, including evaporation, transpiration, condensation, precipitation and run-off (Figure 1.4). Although the total amount of water is fixed, there is concern over the competing demands for its use, availability and quality. Water is a sustainable resource. That means that with recycling and with management, the water can be used again and again. Unfortunately there is increasing human interference in the cycle that can result in environmental **degradation**, conflict and waste.

Since 1900 the population of the planet has increased by 400 per cent, the demand for water in the same period increased by 700 per cent and the amount of water on the planet has increased by 0 per cent.

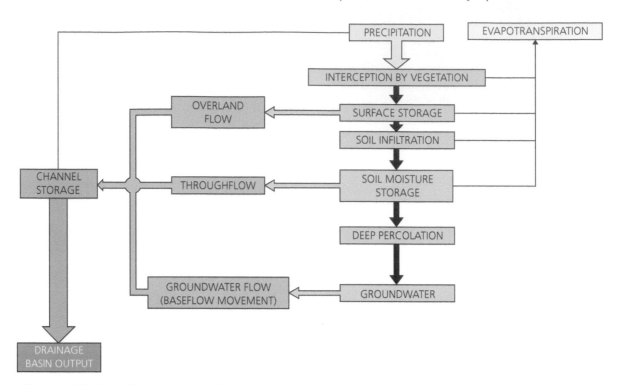

▲ **Figure 1.3** Drainage basin: a system diagram

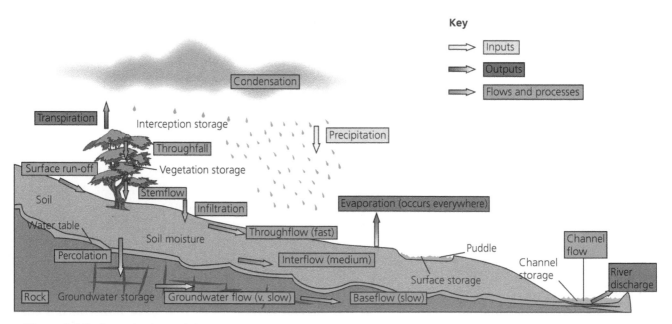

▲ **Figure 1.4** Drainage basin model

Water facts

▲ **Figure 1.5** Planet Earth: water

The points below demonstrate why river basin management is so important.

- If only one-thousandth of the weight of water is from salt, then the water is 'saline'.
- The United States consumes water at twice the rate of other industrialised nations such as the UK and France.
- We take our clean water for granted. However, almost 1.4 billion people do not have access to clean water.

- Each day almost 10,000 children under the age of 5 die as a result of water-borne illnesses in developing countries.
- Over 20 per cent of the world's people must walk at least three hours to fetch water.
- By 2030, 50 countries (with two-thirds of the world's population) are likely to have water shortages.
- The average single-family home uses 100 litres of water each day in the winter and 150 litres in the summer. Consider how you use water in your house. Showering, bathing and using the toilet account for about two-thirds of the average family's water usage.
- Water falls to the Earth as rain, snow or hail.
- When the water reaches the Earth's surface, it may:
  - soak directly into the ground
  - drain into a stream, lake or the ocean
  - evaporate and return to the atmosphere
  - be absorbed by plant roots
  - become part of a glacier or polar ice cap.
- Water that soaks into the ground may be stored as groundwater, or it may slowly move towards a stream. Groundwater may be stored for many years or it may be pumped from a well for household, agricultural or industrial uses. Much of the water found in a stream (when it is not raining or snow is not melting) comes from draining groundwater.
- 0.6 per cent of the Earth's water is groundwater.
- Water contained in lakes, streams and oceans is used for industrial and agricultural purposes.
- Rivers contain 0.002 per cent of the Earth's water.
- Fresh water lakes contain 0.01 per cent of the Earth's water.
- Oceans contain 97.5 per cent of the Earth's water.
- Water is evaporated from bodies of water, the soil and from vegetation. This water vapour forms clouds which can in turn produce precipitation.
- The atmosphere contains 0.001 per cent of the Earth's water.
- Polar ice caps and snow account for 69 per cent of the world's fresh water supply.
- Ice caps and snow contain 2.24 per cent of the Earth's water.
- Since 1990, almost 2 billion people have gained access to a safe water source.

Reflection

A cautionary note about statistics. Someone once said that 80 per cent of all statistics are made up! Or was it 75 per cent or 85 per cent? With the aid of Google, we are not short of information. For example, while researching climate change, I typed in 'flatulent termites as a factor in climate change'. Really?! I had 84,300 hits. The problem is: for all the information at our command, do we really know any more? Statistics can be selected, written and analysed from different perspectives that can be contradictory, biased or simply wrong – so be careful.

Water storage

▲ **Figure 1.6** Ice cap

The water cycle describes the movement of water above, on and through the Earth. This includes water that is locked 'in storage', in other words, water stored in its current state for a reasonable period of time. (In actual fact, more water exists in a stored state than in a moving one at any point through the cycle.) Short-term storage could be days or weeks in a lake, but could be thousands of years for deep frozen groundwater. All of this water is included in the water cycle. Table 1.1 shows the percentage of water stored in the world.

▼ **Table 1.1** Percentage of water stored in the world

Location of stored water	Amount
Oceans	97.5% salt
Other:	*2.5% fresh, of which:*
Ice caps	69% (about 1.7% of all water on the planet)
Ground/soil	30.5%
Rivers/lakes	0.5%

The US Geological Survey (USGS) is the key scientific authority that provides us with data relating to water availability, use, quality and change.

Figure 1.7 shows how the vast amount of water on our planet is held within the oceans and that only 2.5 per cent of all water is stored as fresh water. Of that total, around 69 per cent is locked within ice caps and glaciers. When you consider that 30.5 per cent of the fresh water total is contained within the soil and the rocks underneath, then only a tiny fraction of the total water on this planet is fresh and readily available in rivers and lakes.

Therefore, even though the amount of water locked up in glaciers and ice caps is a small percentage of all water on (and in) the Earth, it represents a large percentage of the total fresh water on Earth.

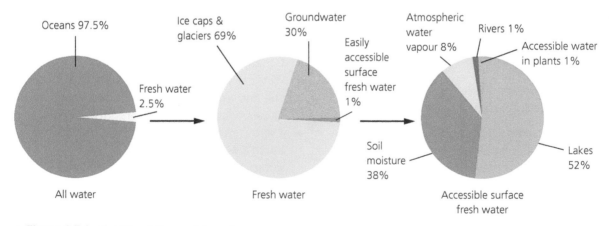

▲ **Figure 1.7** Availability of the world's water

Summary of the hydrological cycle

EVAPORATION is the process of water changing from a liquid into a gas. An input of energy (solar radiation) causes this change

CONDENSATION is the process of water changing from a vapour to a liquid. Condensation takes place when water vapour is cooled

PRECIPITATION
Rain, snow, hail and sleet are the main forms of precipitation

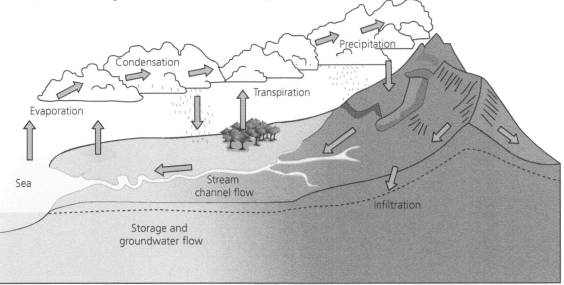

TRANSPIRATION is the process of water loss from plants. Transpiration takes place when the vapour pressure in the air is less than that in the leaf cells, i.e. transpiration is nil when the relative humidity of the air is 100 per cent. In the absence of plant cover, evaporation would still occur from the soil

EVAPOTRANSPIRATION is the combination of evaporation and transpiration from an area

GROUNDWATER is derived mainly from precipitation which has percolated through the soil layers into the zone of saturation, where all pores and cracks are water-filled

STREAM RUN-OFF is the water which finds its way back to the sea via river channels. It is made up of two components: a highly variable storm run-off and a more predictable stable run-off (base flow)

◄ **Figure 1.8**
Hydrological cycle

The balance in a drainage basin is maintained by the hydrological cycle shown in Figure 1.8. The diagram shows the key processes:

- precipitation
- evaporation
- transpiration
- condensation
- infiltration
- run-off.

1.2 Physical characteristics of a selected river basin

You are expected to have sound knowledge and understanding of the physical characteristics of a selected river basin. Let's have a look at the features of a model river basin.

Within the river basin, water erosion, transportation and deposition create the distinctive valley and channel characteristics. Although all rivers and their basins are different, it is possible to identify a model of a typical river divided into three stages: the upper course (or mountain stage), the middle course and the lower course (maturity).

 Task

1 In your own words describe the need for water management.
2 Look again at Figure 1.3 (page 2) and identify the:
   a) key stage phases
   b) flows and processes through the system.
3 'Since 1900 the population of the planet has increased by 400 per cent, the demand for water in the same period increased by 700 per cent and the amount of water on the planet has increased by 0 per cent.' Comment on the significance of this statement.

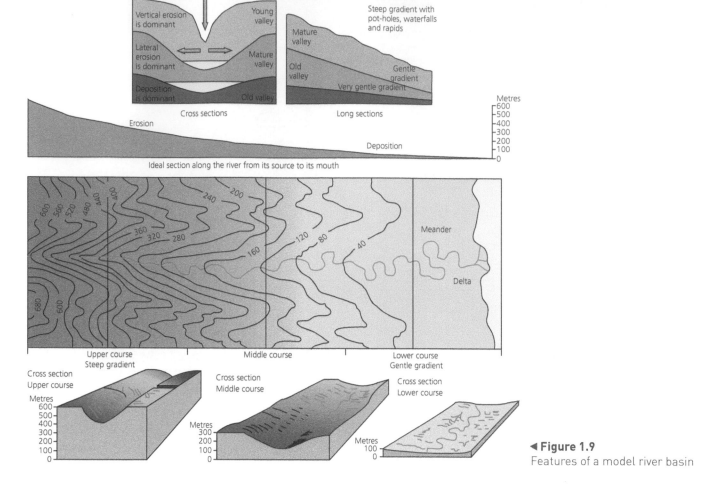

◀ **Figure 1.9**
Features of a model river basin

Global perspective on drainage basins

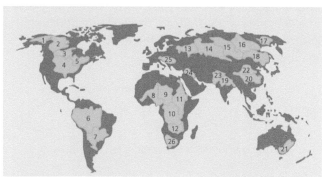

North America	Europe	Asia and Australia
1 Yukon	25 Danube	13 Volga
2 Mackenzie		14 Ob
3 Nelson	**Africa and west Asia**	15 Yenisei
4 Mississippi	8 Niger	16 Lena
5 St Lawrence	9 Lake Chad Basin	17 Kolyma
	10 Congo	18 Amur
South America	11 Nile	19 Ganges and Brahmaputra
6 Amazon	12 Zambezi	20 Yangtze
7 Paraná	26 Orange	21 Murray Darling
	24 Euphrates and Tigris	22 Huang He
		23 Indus

▲ **Figure 1.10** Drainage basins of the world

Almost 50 per cent of the Earth's land surface drains into the Atlantic Ocean.

In North America, water drains towards the Atlantic Ocean through the St Lawrence River from the eastern side of the USA and Canada. The Mississippi River and its tributaries drain a massive area of central USA reaching up into the Canadian provinces (lying between the Rocky Mountains to the west and the Appalachian Mountains to the east), into the Gulf of Mexico.

All of South America east of the Andes Mountains flows east into the Atlantic. The river basins and systems of west and north-west Europe also flow into the Atlantic.

North Africa, southern, central and eastern Europe, and the countries surrounding the Mediterranean have river systems that flow into the Atlantic through the Mediterranean Sea.

Just under 15 per cent of global land surface water drains into the Pacific Ocean. Surrounding the Pacific are the mighty Chinese rivers (including the Yangtze), eastern Russia, Korea, Japan, all the south-east Asian countries, as well as the western side of North and South America.

The Indian Ocean's drainage basin covers a similar area of the Earth's surface (15 per cent). The area covered includes the east of Africa, the Middle East, the Red and Arabian seas. Australia is set within the Indian Ocean's basin area, as is the whole of the Indian sub-continent.

The Arctic Ocean drains just over 15 per cent of the Earth's land surface area. This area includes the north of Alaska, the north of Canada, Scandinavia and large areas of north and interior central Russia.

Finally the remaining balance (around 5 per cent) enters the Southern Ocean from Antarctica.

Of the largest five river basins, two are in South America (the Amazon and the Río de la Plata), two are in Africa (the Nile and the Congo) and one is in North America (the Mississippi). The Amazon, Ganges and Congo rivers contain the greatest volume of water.

A drainage basin with its watershed is the area of land where water will be collected and eventually concentrate as a single body of water (river, lake, reservoir, wetland marsh, estuary, sea or ocean). The watershed marks the point or line where water will flow into one drainage basin or another. It is the 'divide'. From a historical perspective, watersheds played a significant part in the marking and mapping of surface land territorial areas, often being the boundary between countries. Many countries sought to control the complete drainage basin so that conflict with other countries would be reduced. However, the reality was that few countries gained sole control over the basins of some of the world's largest systems. The Mississippi and Yangtze are now each contained within one country, but the Nile, Rhine, Indus and Amazon are within several countries and water disputes are not uncommon. The Aral Sea Basin of central Asia is an example of what can happen when countries get it wrong.

The drainage basin is seen as an important geographical idea in the study of the hydrological cycle and hydrology, the study of water and resources.

North America

▲ **Figure 1.11** Major river basins of North America

So far we have considered a global perspective on drainage basins. If we move to the study of the river basins of North America, we see that there are five key outlets: the Arctic Ocean, Pacific Ocean, Hudson Bay, Atlantic Ocean and the Gulf of Mexico. The sixth area is the Great Basin, with no sea outlet.

North America has a number of major river systems: (clockwise from the north) the Yukon, Mackenzie, St Lawrence, Mississippi/Missouri/Ohio/Tennessee, Rio Grande, Colorado and the Columbia.

Figure 1.11 shows the major river basins of North America. The largest basin, bounded by the Rockies to the west and the Appalachians to the east, includes the mighty Mississippi, Missouri and Tennessee rivers. Rain falling in the eastern side of the Rockies will take a month to reach the sea outlet in the Gulf of Mexico. The western basins of the Colorado, Columbia and Yukon are smaller due to the watershed being far narrower. The Mackenzie system drains northwards to the Arctic Ocean. The large area of central northern Canada (the Canadian Shield) drains into Hudson Bay and then into the Arctic/Atlantic. The main easterly basin covers the Great Lakes and the St Lawrence River, which drains towards the Atlantic.

▲ **Figure 1.12** Major river basins across Africa

▲ **Figure 1.13** Major river basins across Asia

Task

To allow you to consider the background of river basins in a global context, for either North America or Asia, describe and explain the general distribution of river basins.

- selection and development of sites for dams
- need for water management
- consequences of water control projects.

We have seen that water is a valuable resource that requires management. Look at Table 1.2 and consider for a moment why we need a reliable supply of water. How can we control these variables?

▼ **Table 1.2** Users of water and variables

Users	Variables
Domestic	Flooding
Drinking	Drought
Industry	Seasonality
Irrigation	Quality
Power	Competition
Transport	Conflicting demand
Tourism	Sustainability
Recreation	Cost
Environmental	Demand

Task

Table 1.2 shows users of water and variables that can occur. At this early stage in your study of river basin management, identify some of the needs for management and some of the possible conflicts that could occur from the variables.

Water management projects vary in size from small-scale projects to solve an issue in a small river or area to huge projects. Increasingly over the last 50 years a number of massive river basin management projects have been constructed in the developing and the developed world. Projects start with a detailed plan outlining key objectives and outcomes. For example, for the Three Gorges Dam in China, there was a need for flood control and power generation. In the Colorado it was mainly flood control and irrigation for agriculture.

1.3 Control and management of water resources

Our focus of study will now move towards the control and management of water resources, and the following topics in particular:

1.4 Selection and development of sites for dams

When a decision is made to construct a dam, considerable attention and care need to be given to its location. A number of factors, both physical and human, have to be considered, as well as the impact the dam and the resulting reservoir may have on the environment and the activities presently found there.

A dam is a barrier that holds back a river or outlet of a lake; or to put it another way, a dam is a huge wall across a river. Dams are generally constructed to serve the main purpose of storing water, while other structures such as floodgates or levees (also known as dykes) are used to manage the valleys and plains downstream. Dams are often part of a multi-purpose project involving the generation of electricity, navigation and irrigation.

▲ **Figure 1.14** Akosombo Dam on the Volta River, Ghana

An ideal place for building a dam is a narrow part of a deep river valley, because the valley sides act as natural walls. The main purpose of a dam's structure is to fill the valley by holding back all run-off into that area.

One of the largest dams in the world, the Three Gorges, is located in China across the Yangtze. The dam is 2.3 km wide and holds back a 600 km long reservoir. The dam is 185 m high.

As geographers, you should be able to work out the key positive principles when selecting a site for a dam, a reservoir and all necessary infrastructures, as outlined below. No site is perfect – compromise will always have to be negotiated.

Physical factors

A solid foundation is necessary on which to construct the dam. Igneous rocks such as granite or hard metamorphic rocks provide that support. It is best not to build across an earthquake fault line. To prevent seepage from below the dam, underneath the reservoir and the sides of the valley, it is best to make sure that the ground rock is not permeable or porous. In some locations, concrete has been used to line the valley sides to stop any leakage. It is also very important that the bed and walls of the dam should be able to sustain the pressure of the water. In large quantities, water has huge weight and if the walls of the dam are not strong enough to cope, the walls will break and water will spread to the surrounding areas, producing devastating floods that have the potential to cause large-scale destruction of human, animal and plant life.

Costs tend to be lower if the valley is deep and narrow. This would also allow a deep lake to contain the reservoir water, and the smaller surface area would reduce evaporation. Dams across a narrow valley can also have additional strength. It also helps to be in an area with sufficient catchment potential. Reliable and sufficient rain or snow melt within a river basin/catchment area is needed. A cooler climate, with its associated lower evaporation rates, is helpful. So, the ideal is a large catchment area and above average precipitation in order to achieve the greatest potential for storage.

You also need to consider the likelihood of landslides and slope stability and the capability of the dam and reservoir to cope with peak flood flows. If rivers flowing into the reservoir carry large amounts of silt, then there will be an issue over the lake filling up and reducing the efficiency of the dam.

Human factors

If there is a high demand for water or power, then it is more likely that development will take place. Dams are expensive and can be destructive to the existing way of life, so clear benefits have to be seen. Often it is a matter of economic return. In other words, can money be made from the development? You need to consider the impact of the dam and the flooded valley. For example, how much farmland would be lost and how many people would have to be resettled? You also need to consider the impact on any historical sites that would be lost. There will be a need to be sensitive to native cultures. For instance, on the Nile at Aswan many ancient Egyptian temples were lost and along the Three Gorges Dam, 1300 archaeological sites were moved or flooded, several hundred thousand hectares of farmland and orchards were lost, as well as 19 cities inundated. Cost is a factor. Economically developing countries will need to look for external support and financing. You can also calculate the additional land that can now be used for agriculture or the number of people no longer living with the fear of flooding.

Workers (often migrant workers) will be required. You will need to consider the distance to urban and farming communities when hydroelectric power and irrigation are being supplied. Other human factors that need to be considered will be the impact on existing and future communications by road and water.

There will always be an environmental impact on river fisheries, forests, agriculture, settlement, tourism, recreation and wildlife. All dams and their associated developments need to be subject to an **impact assessment**.

Dam failure

Dams do not fail often but, when they do, the damage can be considerable with loss of life and human suffering on a massive scale. In 1975, following extreme levels of rain associated with a typhoon, the failure of the Banqiao Reservoir Dam and over 50 other dams in Henan Province, China, caused more casualties than any other dam failure in history. The disaster killed an estimated 175,000 people and 10 million people lost their homes. In 2012, the town of Campos de Goytacazes in Brazil was badly flooded following a dam failure. The lesson learned from this and other failures is that authorities need to consider their flood warning systems to make sure that people can be evacuated to higher ground before their homes are hit.

The most common causes of dam failure are:

- underlying rocks and geological instability
- constructing the dam in an earthquake zone
- dam failure following poor construction
- very heavy and extreme rainfall
- landslips and the sliding of a mountain into the reservoir (for example, Vajont Dam in Italy, where a massive landslide caused a **tsunami**-type wave to flood over the dam, resulting in 2000 deaths)

- poor maintenance
- dams may be a potential target in times of war and conflict (the Aswan Dam in Egypt, the Three Gorges Dam in China and, most recently, the Mosul Dam in Syria have all been considered to be targets)
- human, computer or design error
- contractors cutting costs by using sub-standard construction materials and techniques.

In conclusion, it is clear that the selection of a site for a dam and its associated features is important and must involve consideration of both physical and human factors. The number of sites that can be developed is limited. Most of the best accessible sites have been taken. Increasingly, new sites are further away from population centres and require lengthy (and expensive) power transmission lines. Over time, such sites may be vulnerable to changes in climate, including variations in precipitation, ground and surface water levels and glacial melt. In a world where many countries are looking to new 'clean' energy, the demand for dams and electricity can only increase.

However, once constructed and if the site is well designed and maintained, a dam as a power source and as part of a managed development can be cheap, environmentally friendly and reliable.

▲ **Figure 1.15** Dam failure: Roosevelt Lake, Lower Colorado; source: US Department of the Interior, Bureau of Reclamation

Power and generation of electricity

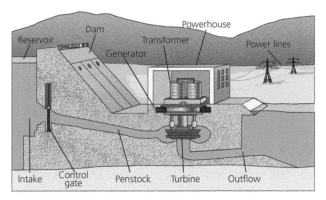

▲ **Figure 1.16** Inside a hydropower plant

In 2018, hydroelectric power (HEP) supplied 22 per cent of the world's electricity and two-thirds of global renewable energy. Much of this is generated by large dams, such as the Three Gorges in China, the Aswan in Egypt or the Itaipu in Brazil/Paraguay. Countries like China, Switzerland and even Scotland use small-scale hydro generation on a wide scale. Most hydroelectric power comes from the potential energy of dammed water; the water may be run through a large pipe called a penstock before reaching turbines. The turbines connect to a generator to produce electricity. A variant on this simple model uses pumped storage to produce electricity to match periods of high and low demand, by moving water between reservoirs at different heights. At times of low electrical demand, excess generation capacity is used to pump water back into the higher reservoir. When there is higher demand, water is released back into the lower reservoir through a turbine (an example is in Cruachan, Scotland, above Loch Awe).

Examination questions for this section will be in the form of the question shown in the task below. Alternatively you may be given a photograph or diagram of a particular case study which you can choose to comment about or refer to another example you have studied.

 Task

Explain the physical and human factors which should be considered when selecting the site for any major dam and its associated reservoir.

1.5 Need for water management

In this section we shall cover the general theory and need for water management.

A report by the World Bank in 2014 provided an overview on the need for and importance of water management. Some of the conclusions of the report are outlined below.

- The world will not be able to meet the great **development** challenges of the twenty-first century.
- The world is facing increased water stress, driven by population and economic growth, land-use changes, increased climate variability and change, and declining groundwater supplies and water quality.
- Water is at the centre of economic and social development: it is vital to maintain health, grow food, manage the environment and create jobs. Water impacts on whether young people are educated or deprived villages can withstand flood or drought. However, mismanagement of this basic element of life has led to millions of deaths and billions of dollars in lost economic growth potential annually, severely restricting a country's development potential.
- Basic sanitation is still unavailable for 2.5 billion people on our planet. Poor sanitation impacts health, education, the environment and industries such as tourism. At least 750 million people lack access to safe drinking water, resulting in 4000 infant deaths each day and yearly economic losses of up to 7 per cent of GDP in some countries.

▲ **Figure 1.17** Children fetching water in economically developing countries

- The global population is growing fast. Analysis suggests that with current practices, the world will face a 40 per cent global shortfall between forecast demand and available supply of water by 2030.
- Feeding a planet of 9 billion people by 2050 will require approximately 50 per cent more water.
- More than half of the world's population now lives in urban areas, with numbers growing fast. How will cities where safe drinking water is already scarce cope with increased demand?
- Currently 2 billion people live in countries with absolute water scarcity and the number is expected to rise to 4.6 billion by 2050.

Source Adapted from an article on www.worldbank.org/en/topic/water/overview

Note: You will have noticed that there are lots of statistics in this chapter. You may well have also noticed that there are inconsistencies in the data. When you research using the web, there are so many different viewpoints. Geographers have to find their information from different sources.

 Reflection

Take some time to consider the importance of the comments above from the World Bank.

Water is a resource that has its limitations. The bulk of our usable water comes from rivers, lakes, the ground, from the soil and falling from the atmosphere. Most people choose to live close to a water supply in order to grow crops and to provide drinking water and energy for factories and our homes. We require reliability and sustainability. Control is often linked to the development of a country, as well as the need and demand for water. Another factor is change. In an area with population or economic growth there will be increasing demand for water. Also, there is no doubt that there appears to be change in our climatic patterns or regimes (the causes of which are tackled elsewhere in this book). Where water is scarce or in greatest demand then the need for management is greatest.

River basin management schemes are often referred to as 'multi-purpose schemes'. As will be shown through a number of case studies below, there are several issues that can be tackled in a river management development.

We shall look at the need for water management in relation to:

- flood control
- power or generation of electricity for homes and industry
- increased food output through irrigation
- a water supply linked to the general raising of living standards and the reduction of **poverty**.

Causes and impact of flooding

The impact of flooding can be grouped under three headings:

- social or human
- economic
- environmental.

Social or human concerns

When flooding occurs, people may die or be injured, and can be made temporarily or permanently homeless, as their property is inundated by water and damaged or destroyed and their possessions may be lost. For many, this is very distressing and has a deep psychological effect. Flood water will be contaminated by sewage and this will lead to a shortage of clean, safe drinking water. Contaminated water places people at significant risk of diseases such as dysentery. Fear of flooding has a negative impact on the confidence of local people.

Economic concerns

Farms, fields, crops and animals can be lost, resulting in loss of income or worse. Crops will be destroyed and can lead to famine, food shortage and increased prices. Businesses often have to be closed and power supplies may be damaged or suspended. Businesses may not recover and jobs and incomes may be lost. Repairs and the cost of rescue have to be paid for and, when insurance is available, premiums will rise for years to come. The infrastructure of the area can be affected, with the loss of roads, bridges and power.

Environmental concerns

Contaminated flood water has a legacy that can last and pollute rivers and flooded land with sewage and rubbish. Animals and their habitats can be destroyed. However, when a river subsides it leaves behind a more fertile landscape formed from the river sediment.

Flood management strategies

- Flooding can cause major human suffering so for centuries we have tried to stop it happening.
- Deaths and destruction from flooding make this 'disaster' the number one in terms of human suffering.
- Flood management is aimed at protecting homes, industry, farming and the environment.
- It has social, economic and environmental impacts.
- We do not have enough money to prevent flooding so we target the most likely areas to suffer and where there is the greatest potential for human suffering. This cost–benefit analysis approach places high value on settlements/factories/communications and fertile farmland.

Hard and soft engineering strategies

Flood management and river basin management strategies generally involve multiple engineering projects that can fall under one of two categories. **Hard engineering** projects are ones that involve the construction of artificial structures that prevent a river from flooding. Engineers and scientists tend to be heavily involved in such strategies. **Soft engineering** projects use natural resources and local knowledge of the river to reduce the risk posed by a flood.

Each type of project has its advantages and disadvantages. Hard engineering projects are generally very successful and have a large impact on the river. This is one of their drawbacks though, as the effects of a hard engineering project can disrupt ecological systems in the drainage basin. Hard engineering techniques generally involve the containment of large volumes of water so, if they were to fail for some reason, the impacts could be many times worse than if the river had been allowed to flood naturally. There is also the high cost, technological requirements and maintenance of hard engineering projects that make them impossible in developing countries without outside aid.

Soft engineering projects focus on reducing the impact of a flood rather than preventing one. Soft engineering projects are significantly cheaper than hard engineering projects, making them more suitable for developing countries. They also have lower education and technology requirements so they can be carried out by local people in remote parts of developing countries.

Soft engineering projects are more sustainable than their hard engineering counterparts. Soft engineering projects are low maintenance and low cost unlike hard engineering projects. In addition, they do not disturb the natural processes and ecological systems in a river basin; instead, they tend to work more with natural processes. Tables 1.3 and 1.4 provide a summary of hard and soft engineering strategies.

Dams create deep reservoirs and can vary the flow of water downstream. This can affect upstream and downstream navigation by altering the river's depth. Deeper water increases or creates freedom of movement for water vessels. Large dams can serve this purpose but most often weirs or locks are used.

▼ **Table 1.3** Hard engineering strategies

Strategy	Key features	Benefits	Drawbacks
Dams	• Dams such as the Hoover Dam are walls built across rivers and their valley. A lake or reservoir is formed behind the dam	• Flood water from upstream will be caught behind the dam, preventing downstream flooding. This water can be controlled through controlled release (often linked to power and irrigation) • The reservoir can be used for recreation and fishing • The dam can be used as a road across a valley	• The largest dams are very expensive • Land is flooded, with consequences for previous land uses (farmland, housing, industry, forest) • Dams trap sediment that once enriched the fertility of the land downstream • Impact on wildlife and fish such as migrating salmon

▼ **Table 1.3** *(Continued)* Hard engineering strategies

Strategy	Key features	Benefits	Drawbacks
Levees or dykes	• Levees are either natural or more likely man-made embankments built along the river	• Allow the floodplain to be developed and reduce (but not eliminate) flooding	• Expensive to construct and maintain. There is always a fear of failure. Levees disrupt natural processes
Channel straightening	• Meanders in the channel are removed by building artificial direct channels	• Flood risk reduced • Easier to navigate, allowing ships and barges to gain access upstream	• This technique has several problems. Flooding becomes more likely downstream of a straightened section of a channel. In addition, erosion is stronger downstream because the river has a lot more kinetic energy • Altering river channels disturbs wildlife habitats • Disrupts natural processes
Diversion spillways	• Diversion spillways are artificial channels that a river can flow into when its discharge rises. These channels move water around an area at risk of flooding and send it either back into the river (but further downstream) or into another river • Spillways are also used to divert water to another river or to a wetland area that can be flooded	• Spillways divert flood water • Spillways generally have floodgates on them that can be used to control the volume of water in the spillway	• Spillways pose a threat to areas near the confluence between the spillway and whichever river it flows into as the discharge here will be increased and so too will the risk of flooding • The path that spillways take can take water around areas not usually used to flooding. If the spillway was to fail for some reason, this could cause widespread damage

▼ **Table 1.4** Soft engineering strategies

Strategy	Key features	Benefits	Drawbacks
Land-use management/ floodplain zoning	• Floodplain zoning involves placing restrictions on land usage (housing/roads) in the areas surrounding a river. It only allows land use for parks, grazing or low-value activities	• If planning does not allow building on a floodplain then the damage caused by the river flooding will be greatly reduced because there is less to damage	• Floodplain zoning limits development to certain areas • If a floodplain has already been developed for housing, it is difficult to get people to relocate off the floodplain • In the UK, we have a shortage of land suitable for housing
River restoration	• Returning a river to a more 'natural' pattern, for example by removing levees	• Allows water to spread over the floodplain • Requires little maintenance and is relatively cheap	• Little impact on the flow of a river and so flooding may not be so easily controlled • Tends to work best in small restoration projects such as at the River Quaggy in south-east England

▼ **Table 1.4** *(Continued)* Soft engineering strategies

Strategy	Key features	Benefits	Drawbacks
Conservation of wetlands and river banks	• Wetlands store and slow down flood waters • Planting shrubs and trees along a river bank, known as 'riparian buffers'	• Reduces downstream flooding • A 'natural' and sustainable scheme • Widens variety of habitats for wildlife, greatly increasing biodiversity	• Reduces the area of land available for farming, which makes this option unpopular among farmers
Afforestation	• Involves the planting of trees in a drainage basin to increase interception and storage while reducing surface run-off	• Reduces a river's discharge and so makes it less likely to flood • Prevents mass wasting which reduces the amount of soil entering the river and keeps the river's capacity high • Creates new habitats for animals and can improve water quality by filtering pollutants out of rainwater	• May be conflict from farmers, tourists and other land users
Monitoring strategies	• In the UK, the Environment Agency monitors weather forecasts, river discharge, tidal surges and other sources of information • Information is disseminated via newspapers, TV, radio, the internet and text messages • Environment Agency co-ordinates appropriate responses	• Warning people enables them to take appropriate action regarding flooding • Lives can be saved and damage reduced	• Cannot prevent incidents from taking place • Sometimes warnings have been given that have not resulted in a flood situation; this has meant that some people adopt a casual attitude and assume that every warning is an over-reaction • Not everyone has access to the information being given out • A flood warning in Bangladesh is going to be more difficult to convey to people than a warning about the River Thames

Tackling exam questions

The river basin management exam question may ask you to study a variety of maps, tables, graphs and diagrams and then comment on and explain the need for water management. Usually a specific example will be given. It may be the Mississippi, Colorado, Nile or indeed any basin, but how are you expected to answer a question about a river basin that you have not studied? You could compare it to driving a car along an unfamiliar road: once you learn the key skills you can then apply the key principles and skills in any situation.

Remember that in Geography at Higher level you are expected to have detailed knowledge of not only the theory but also some specific, named case studies.

So, what are the key principles for answering a basic examination question on river basin management projects?

- Check total rainfall. Study the visuals. You may be given a graph with two or more sets of climate data. Check seasonal imbalance and/or unreliability evidence.
- Check for the possibility and/or need for transfer of water from an area with a water surplus to a deficit area (by aqueduct/canal or river).

- Be aware of the need to reduce the likelihood of flooding from seasonal precipitation and/or snow melt from any surrounding mountains.
- What evidence is given of water usage in the area? There could be mention of population growth, big cities, industry, tourism and agriculture. There will be a need to provide a regular water supply.
- The need to generate energy (HEP) for industry, farms, communities.
- Generally in a less developed country there will always be the need to raise the standard of living and help to reduce poverty.
- If the river is large enough, then there could be the need to improve navigation potential.
- Is there evidence that suggests more than one country or state could be involved?

1.6 Consequences of large-scale water management projects

River basin water control projects change the hydrological cycle of a river basin. Some of the effects of large-scale water management projects include:

- changes in storage
- changes to evaporation and condensation rates
- construction of a barrier or dam
- channelling of rivers
- controlling the level of rivers and reservoirs
- draining and irrigating the land
- surface run-off altered
- less water below dam
- lakes which create their own micro-climates
- infiltration rates into ground affected
- seasonal variations in river levels altered
- changes in the level of the water table
- less water flowing into the sea
- changes in the salinity of the water and land.

Case study: the Three Gorges, China

The Three Gorges Dam and water management project is the largest project in China ... at least since the construction of the Great Wall.

> 'He who controls the water, controls the people' – a Chinese saying

When researching this case study, I was amazed at the wide variety of 'facts' and different claims that were made. The message for you is that sometimes the 'evidence' can be contested, disputed and even exaggerated or falsified. Double check and take nothing at face value.

This case study of the Three Gorges is a very detailed review, and from the information given you should be able to answer the questions on the key examination themes given at the end of the case study.

Introduction

The Three Gorges Dam sits across the Yangtze River in China in the province of Hubei.

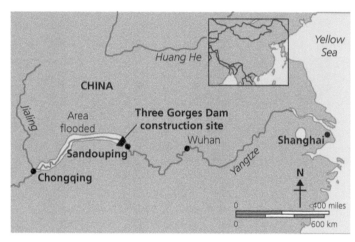

▲ **Figure 1.18** Location map of the Three Gorges Dam and Reservoir

For those of you who love statistics, the Three Gorges Dam consists of a main dam wall, a series of five locks and 34 generators to produce electricity. It is the world's largest hydroelectric dam (based on generating capacity). The dam is 2.3 km long and over 180 m high. At its base it is 115 m thick. Construction involved 27 million m³ of concrete and 0.5 million tonnes of steel.

The reservoir is 600 km long and 91 m at its deepest point. The cost of construction varies from $30 billion to $59 billion and is expected to have paid for itself through electricity sales by 2025. The reservoir helps control flooding on the Yangtze River Basin and allows 10,000-tonne ocean freighters to sail into the interior of China (six months of the year). The 32 main turbines are capable of generating 3 per cent of China's electricity. The dam took 15 years to construct.

History

The dam was first proposed in 1919 by Dr Sun Yat-Sen. Throughout the 1930s plans were drawn up by the Japanese, the Chinese and, by the 1940s, the Americans. All of this was abandoned due to the Chinese Civil War and internal conflicts. Following flooding in 1954, when over 33,000 people died, talks started again. More than 330,000 people have died in three massive floods over the last 200 years. More delays followed the 'Cultural Revolution' and the 'Great Leap Forward' but, by 1994, some 75 years since it was first proposed, the construction of the Three Gorges Dam finally began. The dam was operational in 2009 but continuous adjustments and additional projects are still ongoing. Apart from the Three Gorges project, there are many other dams along the river, with at least 46 projects either completed or under construction.

Winding about 6379 km, the Yangtze River is the largest in China and the third largest in the world after the Nile and the Amazon. Originating from the Tanggula Range in western China, it crosses 11 provinces and cities from west to east, and finally enters the East China Sea at Shanghai. It has numerous tributaries including Min River, Han River, Jialing River, Gan River and Huangpu River. The Three Gorges Dam is the largest dam project and hydroelectric power station in the world. Generally, people consider the river a dividing line between north and south China. Areas to the north and the south of the river have many differences in climate, scenery, economics, culture and folk customs.

Dr Sun Yat-Sen in 1919 mentioned the need to dam the river to 'control flooding and generate electricity'. So 100 years later, does this still apply today?

Need for the Three Gorges project

The key functions of this project are flood control, power generation and improved navigation.

Seasonal flooding was common around the Yangtze. Over many parts of the basin, the rainy season (June–August) resulted in a massive increase in the river's discharge with accompanying flooding. In 1954, the flood killed 33,000 people and inundated the city of Wuhan for almost 100 days; 18 million people were displaced.

It is a challenge to establish facts about China's demand for power. Statistics indicate an output growth of over 10 per cent per year in power for the last 15 years. There has been a doubling of demand over the last 10 years, from all sectors of the economy. Leading the surge is a booming industrial sector. At present 69 per cent of power comes from fossil fuels (coal, gas and oil) with 6 per cent from wind and tiny amounts from nuclear and solar. Hydro sources remain around 23 per cent and the output from the Three Gorges is very significant at 3 per cent of China's needs.

The river traditionally had limited navigation use because of fluctuating water levels and flooding. Two sets of canal locks were built, each set with five steps or stages. Water navigation for both people and goods has resulted in a major increase in water transport. Navigation is also far safer, since the Gorge was notoriously dangerous and hazardous to navigate – exciting and spectacular for tourists, but overall fraught with danger.

There have been five major floods in the last 100 years. Flood protection is mostly achieved through what is called 'hard engineering' defences. These are man-made structures that reduce flooding. The most common types of hard engineering techniques are dams, channel straightening, levees and diversion spillways.

At this site the dam is clearly the most significant hard engineering factor. Along the course of the Yangtze River over 3600 km of levees have been constructed. Levees are reinforced raised banks along the side of the river.

Figure 1.20 shows how levees can also be a natural consequence of flooding. However, along the Yangtze man-made levees (also known as dykes) have been constructed. Do they work? Yes (and, because this is Geography ... no!). Levees do hold back flood water. However, if the river is exceptionally high, water can either flow over the levee or possibly break through.

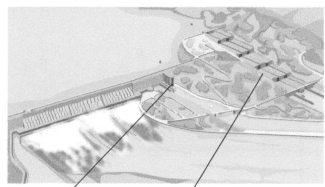

Ship lifting tower – a large and powerful elevator for ships less than 25 m long – faster than using the five-level lock

Five-level double ship lock – 1600 m long, this lock can raise or lower ships a total vertical distance of 113 m. It is the largest lock system in the world

▲ **Figure 1.19** Sketch of the site of the Three Gorges Dam

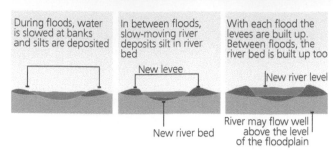

During floods, water is slowed at banks and silts are deposited

In between floods, slow-moving river deposits silt in river bed

New levee

With each flood the levees are built up. Between floods, the river bed is built up too

New river level

New river bed

River may flow well above the level of the floodplain

▲ **Figure 1.20** Levees and flood protection

Since the top of the levee is above the level of the surrounding land, the flood waters now spread rapidly over the valley floor. In 1998, floods breached the levees. The levees were reinforced and raised and were effective at reducing the impact of the floods in 2002. Since then, China has continued to strengthen and raise the levees. The chance of flooding has been greatly reduced downstream from the Three Gorges Dam. However, history has taught us that the exceptional flood could happen in the future. Levees allow the floodplain to be developed for industry, housing and farming. However, they tend to be expensive to construct and maintain and there always remains the risk of breaching.

Positive consequences of the Three Gorges Dam and management project

▲ **Figure 1.21** General view of the Three Gorges Dam and lake

'A triumph of human determination and ingenuity?'

Source Steven Mufson, *Washington Post*, 9 November 1997

The dam now helps to protect cities such as Wuhan, Nanjing and Shanghai from flooding. Lots of farms, villages, towns and factories are also located near the river. There are 22 km² of space behind the dam to hold back the flood waters. In the dry season the water is released down the river and the reservoir level drops to prepare for the next flood.

Goods and people can now navigate freely along a considerable length of the Yangtze. River traffic is up six times, with a 35 per cent decrease in shipping costs.

In the past, raw sewage mainly entered the river system. Now over 65 per cent of sewage is treated in over 50 water treatment plants and solid waste is increasingly collected in landfills.

The government of China could be seen to be actively attempting to reduce their levels of pollution. This is a useful propaganda issue at a time when China is regularly criticised in the media for its power and industrial levels of pollution.

The 34 generators produce some 22,250 megawatts of electricity. This is an impressive amount, some 3 per cent of China's needs. Homes, factories and farms will have a regular and reliable source of electricity.

It is said that there are positive environmental green issues. The dam has saved some 450 million tonnes of coal being mined and burned per year. The 'green' energy also reduced emissions (for example, 100 million tonnes of **greenhouse gases**, 1 million tonnes of sulphur dioxide and 10,000 tonnes of carbon monoxide per year). China still relies on coal power stations for 70 per cent of its energy, but projects such as this reduce reliance on fossil sources. The new planned villages and towns were to have better houses and new schools and hospitals. An agreement with the authorities was that every person displaced would have improvements in all aspects of their living conditions. It was expected to reduce levels of poverty. There is evidence to suggest that this did happen, but possibly not for 'every person'.

Negative consequences of the Three Gorges Dam and management project

'A mammoth folly, a triumph of ego and political showmanship'

Source Steven Mufson, *Washington Post*, 9 November 1997

People had to relocate as the water level in the reservoir lake rose. Possibly up to 2 million people in total were 'forced' to move, with 13 cities and 1352 villages submerged. This compulsory relocation has not been viewed well by many of these people. The new sites have led to deforestation and soil erosion.

▲ **Figure 1.22** Buildings waiting to be submerged by the dam

Several hundred thousands of hectares of farmland were lost and 1660 factories and 1300 sites of cultural and historic interest flooded, including the temple of Zhang Fei. The new land is not as fertile as the lost land, often being steep and prone to landslides and erosion. Neolithic settlements and ancient cultures will be lost to scientific research. Some of the historic sites have been moved to higher ground. Although the gorge remains scenically awesome, the flooding of the valley has meant that some of the most spectacular limestone gorges have been lost. Possibly there will be a reduction in tourism to this area.

Although the threat of flooding has been reduced massively, it has not been eliminated. There has been increased flooding from some of the tributary rivers that now flow into the reservoir. One well-known example of this is along the Daning River, which not only has flooded but there is now increased erosion of the river bank causing landslips and collapse.

The rich alluvial sediment is now trapped behind the dam, not only resulting in decreased fertility downstream but also causing the silt to be trapped, raising the bed of the river behind the dam. The sediment build-up in the reservoir has altered or destroyed floodplains, river deltas, ocean estuaries, beaches and wetlands, which provide habitation for spawning fish. Since the silt does not go downstream any more, the city of Shanghai will suffer because there has been no silt to replace land that is washed away at the river's mouth. The silt can also clog the dam's outlets, preventing the release of water. If left, the dam will no longer hold back the water, allowing the water to flow over the dam, causing it to collapse.

Early research indicates a negative reduction in the number of birds and animals, such as the Siberian crane, Yangtze finless porpoise and Baiji dolphins. This basin has long been regarded as a biodiversity 'hotspot', being home to over 6400 plant species, 3400 insect species, 300 fish species and 500 vertebrates.

The reservoir covered old factories, mines, hospitals, rubbish dumps and graveyards. There have been increased levels of toxins in the water system, such as arsenic, sulphides, cyanides and mercury.

The dam was originally designed to supply 10 per cent of China's electricity requirements. The failure to hit this target is due to the growth in demand being far greater than expected.

This development will not be the last. At least four other major projects are planned or being constructed and over 40 smaller projects.

Employment after relocation

Forty per cent of people living in the Three Gorges area are farmers. These people were promised compensation for their land by the government, but the problem was that some of the land that had been identified for these people was considered to be unusable. In fact, only 10 per cent of the new available farmland was arable. The government had only addressed the issue of land compensation for 60 per cent of farmers that had been forced to leave their land. The remaining farmers were forced to find urban jobs and in effect, to change their entire lifestyle. The most pressing problem was that these rural people were often thrown into a job that they knew nothing about. China's rural unemployment rate in the province was high so the likelihood of finding jobs was slim. Once all of the people of Three Gorges had been relocated, the unemployment rate rose with thousands of people forced to live in poverty. There has been considerable migration to cities such as Shanghai.

Although not, perhaps, one of the most pressing issues in the current debate over the Three Gorges Dam, disease has become a concern of many biologists. The dam basin has a diameter of several kilometres, and this massive pool of partially standing water is a perfect breeding ground for malaria, snail fever and other water-related diseases. In 1996 there was an outbreak of malaria in the Three Gorges area, and the dam basin could severely increase the chance that it will happen again.

Three Gorges resettlement

- Thirteen major cities, 140 smaller cities and towns, around 1350 villages, 1600 factories and 700 schools were submerged by the Three Gorges project. Wanzhou was the largest victim. Two-thirds of the city, including 22 km² and 900 factories, were submerged. In compensation, the new city of Wanzhou had a new railroad, a new highway linking it to Shanghai and a new mountain-top airport that can handle jumbo jets. Remember that this development brings the opportunity for real industrial growth to an area that was relatively poor through isolation, lack of power and poor infrastructure.
- Low-lying Yunyang was also hit hard. More than 160,000 people from the town had to move and countless numbers of buildings were submerged. Before the waters in the reservoir began to rise, areas that were submerged were stripped of anything that could be sold. Some places looked like they had been bombed. Thirteen replacement cities are currently being built along 595 km of

waterways affected by the dam. For example, New Yunyang is named after the city that had been submerged, and New Zigui was built on a scenic promontory selected with tourism in mind. The Jiangdu Temple was moved there.

- About 1.4 million people were relocated to make room for the reservoir created by the Three Gorges project and they have been relocated all over China. Some have been sent to the Shanghai area, while others have been sent to Guangzhou. Some have also been sent to Tibet and other remote places. Many have been resettled in new communities near their old home towns. Some remained in what was left of the old towns.
- In many ways relocating so many people has proven to be a more daunting task than building the dam itself. The government allocated $10 billion for relocation, about 40 per cent of the cost of the Three Gorges Dam project. People were often moved in blocks. Sometimes entire hamlets were loaded onto a boat and sent downstream, where the government provided the villagers with new plots of land. Many relocated people were happy about the move. They left behind a mud-walled, dirt-floor hut with an outdoor toilet, often with no windows and either no or unreliable supplies of electricity, clean water and a sewage link.
- The new apartments were built to a relatively decent standard with water, gas, electricity, toilets and rents of less than $2 a month. Some people earned enough in compensation to buy a couple of houses and rent them out for income.
- The last town, Gaoyang in Hubei province, was evacuated in July 2008, allowing the reservoir to reach its final height of 175 m above sea level. The 1000 or so households in the town were relocated.
- Critics have complained that the government has fallen far short of its goals in helping to resettle the 1.5 million people displaced by the rising waters behind the dam. Relocated people have complained about inadequate compensation, a shortage of jobs, and corruption that robbed them of money they were entitled to. One woman looking back on her forced relocation said, 'We had no idea when we were going to be moved. They suddenly turned up one day and told us to get our belongings

together. You had to take what the state gave you. There was no bargaining.' The government position was summed up by the slogan: 'Forsake the small home. Support the big home.'

- Some families were forced to abandon homes their ancestors had occupied for generations. In addition to moving their belongings, many displaced people also wanted to move the graves of deceased loved ones. After they left, their houses were torn down to discourage people from moving back.

- Some people who were promised $4000 a head only received $1000. They were unaware of deductions for moving, down payments on the new homes and a variety of other fees. Other people received no compensation or explanation of what happened to their money.

- The people who had the hardest time were the ones with no connections to work units, which are the main channel in communist China for distributing social benefits and exerting social control. Those without work units were unable to get compensation, new housing or other benefits. Migrants without residency permits for the area also had similar problems.

> 'The long-controversial Three Gorges Dam project is expected to bring serious environmental harm to the Yangtze River estuary, and Shanghai will bear the worst of it. According to Professor Chen Guojie, the project is going to reduce the amount of river water and sediment in the Shanghai area and sea water will encroach on and cause erosion to the coastline. "For Shanghai, what is good and useful is decreasing, while the harmful and toxic is increasing. Lower water levels in the river could lead to the intrusion of sea water, which would impact the fish ecology in the area."'

Source Adapted from article in *South China Morning Post* by Mandy Zuo, 17 July 2014

While in Shanghai, I spoke with a student who explained how the relocation had had a positive effect on some people:

> 'Mr Chan and his wife and child, his brother and wife with their child, his mother and aunt were relocated from their village outside Gaoyang. He was a farmer, struggling to make any surplus from his farm. His house was lacking any connection with mains electricity, or water and sanitation. The local school was a more modern building as was the health clinic. Communications into Gaoyang were poor and unreliable. The whole family were happy to be relocated.
>
> Their new house was clean, had power and sanitation. The rent was higher than he had been told. The roads were paved and the transport network cheap and reliable. There were no farming jobs so he was happy to work as a security guard and his wife found employment in a clothing factory. Their income was double that from before. His brother chose to migrate to Shanghai. He moved without the necessary permit, but had found well paid employment as a construction worker. Asked if he was happy, Mr Chan replied, "I was sorry to leave my friends but we are now building a new life here."'

▲ **Figure 1.23** Police guarding the Three Gorges Dam

▼ **Table 1.5** Summary of the Three Gorges Dam project

Benefits/Advantages	Drawbacks/Disadvantages
Social	**Social**
• End of the most serious flooding with 15 million people downstream free from the threat of flooding • Will improve the quality of life for millions with electricity in homes and farms • People will be compensated for the loss of their homes and farmland • Promise of better house, better jobs and improvements in health and education • Should help to curb rural depopulation from the basin	• 13 cities, 140 towns and around 1350 villages had to be flooded • Up to 2 million people resettled, causing massive social and human upheaval • Broken promises regarding compensation and quality of resettlement • Human rights issues over uncompromising approach • Communities and families torn apart following resettlement
Economic	**Economic**
• 34 generators, 22 megawatts of electricity for homes, factories and farms • Shipping tonnage increased from 10 million to nearer 100 million • Overall transport costs reduced by 35% • Ships of over 3000 tonnes able to reach 2400 km from the coast • Jobs and a money economy to boost output, promote trade and increase overall wealth of the region	• Uncertainty over the costs, which vary from \$22.5 to \$59 billion • Loss of over 1650 factories and good fertile soil • Massive loss of existing jobs in textiles, engineering and food factories • New jobs not often available in new sites • Massive loss of tourism revenue. Will fewer people come to the Three Gorges? • Estimates vary over the time it will take to cover development costs (from 20 to 50 years)
Environmental	**Environmental**
• Massive saving in use of coal • Large reduction in greenhouse gas emissions creating a clean, environmentally friendly area • New reclaimed areas with new habitats for birds and wildlife • River and lake water cleaner and safer for navigation • Afforestation on the river banks and surrounding slopes	• High levels of sewage in tributary rivers • Water pollution in lake and downstream following flooding of mines, landfill sites and factories • Damage to existing habitats of fish, animals and waterfowl • 1300 archaeological sites flooded or moved • Earthquake zone fear • 530 million tonnes of silt per year will clog up dam • Flooding will not be eliminated since rivers can never really be controlled • Increased bank erosion downstream of dam • Dredging required in reservoir and rivers • Possibility of water-related diseases such as malaria
Political	**Political**
• Massive project is symbolic signature to the world that China is an international power • Government proud of the scale of this project • Has helped to encourage trade and financial links to the rest of the world • Position of Shanghai as a global finance centre increased • Has brought more remote inaccessible parts of China closer to the power in the East • Reduces the chaos and political fury that would erupt if there was a repeat of the disastrous flooding of the past	• Considerable arguments between the Chinese Government and local groups, environment groups, tourism and human rights groups • Opposition often suppressed and little freedom to complain • Corruption cases now being exposed • Concern that the dam could be a target for terrorists

Cautionary note

In my research for this case study I consulted textbooks, articles in magazines, newspapers and general internet search engines. I visited China and observed the Gorge for myself. I was a tourist on the river. There was no shortage of information. So how do you value and balance the conflicting information? There is bias, not only from politicians in China, from village people forced to move their homes and from a hostile world outside China. As a geographer you will need to judge all this evidence yourself. Depending on your perspective, you will come to a decision. I am aware that there does seem to be a powerful negative balance presented in Table 1.5, but the reality is that the Three Gorges does exist and we have to deal with the issues that arise from it.

Research opportunity

Now conduct some research of your own. See if you can answer the following questions:
- What are the physical characteristics of the Yangtze River Basin?
- Explain the human and physical factors that had to be considered when selecting the site of the dam and reservoir.

Task

Having read through the information on the Three Gorges Dam project and conducted your own research, you should now be able to answer these questions:

1 Explain the need for water management on the Yangtze River.
2 Describe and explain the social, economic, environmental and political benefits and adverse consequences of the project.

Case study: the Colorado River Management Basin and the Hoover Dam

Site of the Hoover Dam

The Hoover Dam is a massive concrete dam in the Black Canyon of the Colorado River. Initially named the Boulder Dam, it is sited on the Arizona and Nevada border. The project took just under six years to complete (in 1936), and was funded by the US Government. The 1930s were hard years in the USA (the Great Depression) and construction provided employment for many thousands of workers. During construction, 120 lives were lost.

The Black Canyon area had previously been surveyed as suitable for such a dam. The project was multi-purpose in that the dam would control the seasonal flood waters, the lake would provide water for irrigating farmland and the associated power station would generate enough electricity to revitalise the region and states around.

At the time of planning this was the largest concrete dam that had ever been constructed and a group of six companies joined forces to make the successful bid. Originally an eight-year project, it was completed two years ahead of schedule.

▲ **Figure 1.24** Hoover Dam under construction

This was never going to be an easy construction with extreme temperatures during the summer, the relative remoteness of the site and a lack of nearby roads and settlement. The Hoover Dam holds back Lake Mead. The dam is sited 50 km south-east of Las Vegas. Since the late 1930s, the generators have provided power for industry, homes and farms.

Today, the area around the Hoover Dam is a major tourist attraction, with over 6 million visitors and over 1 million people visiting the dam site and museum. Highway 93 also runs across the dam, providing an important route across the area. In 2010, a new highway and spectacular bridge was opened, making it easier for even more visitors to come to the area.

Colorado River Basin

The 2330 km Colorado River is the principal river of south-western USA and north-west Mexico. The basin covers seven US and two Mexican states. Much of the area suffers from low precipitation so the Colorado and its dam is a major resource. The Colorado rises in the central Rocky Mountains and flows south-west across the Colorado Plateau until it reaches Lake Mead. Here it turns south towards the Mexican border where it forms a large delta at the Gulf of California.

The Colorado River Compact was an agreement reached in 1922 between the seven US states in the basin of the Colorado River, which established the rules (protocols) governing the allocation of water.

With so many interested parties, it is not surprising that the decisions taken and the outcomes remain contested. Each state wished that their voice was heard, and on many occasions the Federal Government had to step in and enforce compromise to make sure that progress could be made.

The Colorado is a major resource. This can be seen in its extensive use for irrigation (agricultural) and urban water in the desert lands within the basin. Like all great resources, it needs to be managed so that the impact is co-ordinated and it operates at best efficiency to reach its full potential. There is an extensive system of water storage and transfer using dams, reservoirs and aqueducts which transfer water

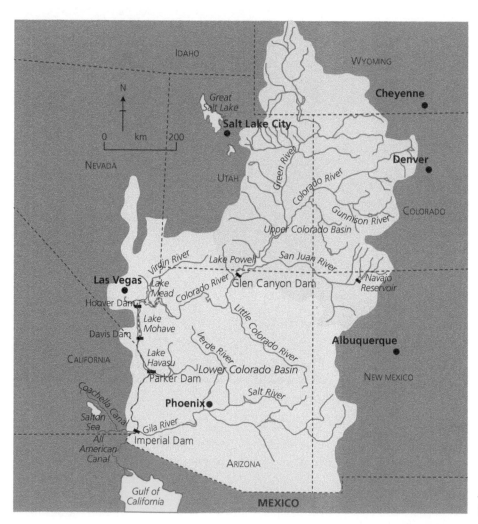

◀ **Figure 1.25**
Map of the Colorado Basin

for irrigation and industrial and domestic water supply to almost 50 million people, both within the basin and around it.

The underlying geology provided the solid base for the dams. The impermeable rocks were perfect for water storage in the large lakes that would be created. This, alongside the river's large flow and steep gradient, makes it perfect for the generation of hydroelectric power. However, over the last 50 years, such has been the demand for water and a pattern of lower than average precipitation that the river in its lower course now barely exists. For most of the year there is no more than a trickle flowing over the final 160 km.

As can be seen from Figure 1.25, the basin has several dams, including the Imperial, Parker, Hoover, Davis and Glen Canyon, and reservoir lakes such as Lakes Havasu, Mohave, Mead, Powell and Navajo.

The Hoover Dam is 270 metres from the base rock to the roadway at the top of the dam. Someone has calculated that it weighs almost 7 million tonnes! The pouring of the concrete was almost continuous for two years, completed in May 1935. During construction four diversion tunnels were excavated, which allowed the Colorado water to pass around the dam site.

The foundation and valley sides are rock of volcanic origin, geologically known as 'andesite breccia'. The rock is hard, durable and provides a very secure base. The site provides the ideal set of factors for the construction of such a massive dam being solid, impermeable, hard-wearing and geologically stable, with no significant history of tectonic activity.

During construction a total of 22,000 men worked on the dam. The work was hard and dangerous. The environment is arid with extreme temperatures during the summer months. There was little in the way of existing infrastructure (roads or power) in the area, so a new town, Boulder City, was required to house both government and contractor workers. Power and communication lines and a new surfaced highway from Boulder City to the dam site were required. Over 80 km of railway track was needed, together with a new 240 km power transmission line from San Bernardino, California, to the dam site to supply energy for construction. Of course, once the dam was constructed, the electricity was carried in the other direction. The dam cost $50 million.

The valley is spectacular, with steep valley sides and a deep site suitable for the new lake. The catchment of the mighty River Colorado was vast, with precipitation falling over the mountains. There was a high degree of variability in water supply. Since this is an area of high temperatures, evaporation rates are high, although the nature of the lake – relatively narrow and deep – did help to reduce evaporation losses. However, not all sites are free from negative issues. Around Lake Powell there are sedimentary rocks and the lake loses 300 km^3 annually due to seepage. In some locations concrete has been used to line the valley sides. The river flow is in a largely desert area, much of it receiving less than 250 mm of rainfall per year. Although rainfall is both unreliable and seasonal, there is a real possibility of thunderstorms and flash flooding. Lake Powell took almost 18 years to fill, finally reaching capacity in 1980, with a maximum storage capacity of 35 km^3.

There is a large market for the power, irrigation and drinking/industrial water for some 44 million people and the cities of Las Vegas, Phoenix and Los Angeles.

The river experiences wet and dry periods that last for years; it is currently going through a drier period. However, 1983–87 were relatively wet years with a reliable and more consistent river flow. Climate projections show that predicted higher temperatures as a result of global climate change are likely to result in shorter winters, earlier spring run-off, increased evaporation and a reduction in the Colorado River stream flow and water supplies.

 Task

What are the main physical characteristics of the Colorado River Basin?

Need for the River Basin Management programme

The Colorado River Basin Management project focused on:

- flood control
- electricity from hydroelectric power to factories, farms and houses

- drinking and domestic water for south-west USA, therefore supporting the huge population growth in the region (see Table 1.6)
- irrigation water for the potentially rich agricultural areas, thereby increasing food production.

The Hoover Dam did, and continues to do, all the things its supporters hoped it would. It protects southern California and Arizona from the disastrous floods for which the Colorado had been famous. It provides water to irrigate farm fields. It supplies water and power to Los Angeles and other rapidly growing cities in the south-west. But the dam also had an entirely unexpected result, one that began while it was still under construction. For millions of people in the 1930s, including those who would never visit it, the Hoover Dam came to symbolise what American industry and American workers could do, even in the depths of the Great Depression. In the early twenty-first century, almost a million people still come to visit the huge dam every year, and many millions more visit the wider area.

Task

Evaluate the benefits and adverse impacts of the Colorado River Basin Management programme, under the headings:

- Social
- Economic
- Environmental
- Political.

'We are on the verge of a water crisis. By 2025, more than half of the nations in the world will face freshwater stress or shortages and, by 2050, as much as 75 per cent of the world's population could face freshwater scarcity. International Alert has identified 46 countries, home to 2.7 billion people, where climate change and water-related crises create a high risk of violent conflict. A further 56 countries, representing another 1.2 billion people, are at high risk of political instability. That's more than half the world.'

Source www.waterpolitics.com (an excellent source for your own research)

'I believe water will be the defining crisis of our century, the main vehicle through which climate change will be felt from droughts, storms, and floods to degrading water quality. We'll see major conflicts over water; water refugees. We inhabit a water planet, and unless we protect, manage, and restore that resource, the future will be a very different place from the one we imagine today.'

Source Environmentalist Alexandra Cousteau

Water politics and the Berlin Rules

Rivers flow across international borders. Often a river basin will cover several countries. Consider the Rhine (Switzerland, France, Germany, Netherlands), the Colorado (USA, Mexico) and Indus (Pakistan, India).

How do we balance the resource so that there is fairness? Countries do compete for water resources, so there are international treaties and protocols regarding how to manage and share those supplies. As political situations change, then pacts can disintegrate and major disputes arise. The power usually lies with the upstream countries who control outflow.

In 2004 the *Berlin Rules on Water Resources* reinforced and updated the internationally accepted Helsinki Rules (1966) regulating how rivers and their connected groundwaters may be used when they cross national boundaries. This is now the accepted protocol for such

▼ **Table 1.6** Population growth in Phoenix and Las Vegas

City (metropolitan area)	1990 population	2000 population	2010 population	2020 population (estimate)	% population change (1990–2010)
Phoenix	2,300,000	3,300,000	4,300,000	6,500,000	+87%
Las Vegas	800,000	1,400,000	1,900,000	2,600,000	+137%

Source US Census Bureau, 2010

matters although there are no formal mechanisms to enforce the guidelines. The Berlin Rules support the rights of all bordering nations to have an 'equitable share in the water resources', while taking into consideration 'past uses of the resource and balancing the needs and demands of the bordering nations'.

The Berlin Rules for water sharing take into consideration the following:

- natural factors: rainfall amount, sources of water and share of the drainage basin
- efficiency: avoidance of waste and mismanagement
- social needs: population size and development
- economic needs: maintaining economic viability and **sustainable development** of surrounding nations
- impact downstream: pollution, lower water table, insufficient water left in system
- dependency: are there other water sources/river basins?
- prior use: balancing previous demands with future demands.

What is the reality? Agreements are never 'fair' and usually the country with the greatest economic, military or political power gets the better deal. Take the Colorado River and the relative wealth and power of Mexico and the USA, for example. Reaching an agreement can be complicated and can lead to serious disputes, for example between Pakistan and India.

Politics and the Colorado

In the Colorado, water rights relating to the states date back to 1922. Since then, the politicians have

signed treaty after treaty; today there are seven states and two countries involved.

These 'Laws of the River' established the division of water between the upper basin states (New Mexico, Utah, Colorado and Wyoming) and the lower states (California, Arizona and Nevada) and Mexico.

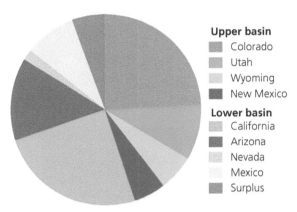

Upper basin
- Colorado
- Utah
- Wyoming
- New Mexico

Lower basin
- California
- Arizona
- Nevada
- Mexico
- Surplus

▲ **Figure 1.26** Water allocation from the Colorado River

California obtained (through its size, economic and political power) the largest proportion of the water. This has been reduced through the courts and new developments elsewhere. In the 1920s rainfall was higher (on average) and more regular than today. Assumptions were made about these high levels and, combined with increasing demand from population growth, industry and farming, demand and use was higher than supply.

Figure 1.27 shows the long-term drop into Lake Powell. It is perfectly reasonable to assume that inflow will pick up again, although the trend is worryingly down.

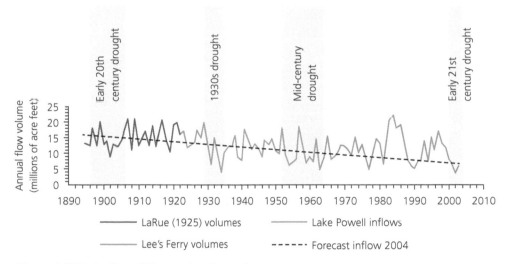

▲ **Figure 1.27** Lake Powell fluctuating flow volume

Competing demands and conflicts between the key users of the Colorado River Basin

▲ **Figure 1.28** The need for water management creates conflict around the globe. Above are contrasting examples from Malaysia (top) and Louisiana, USA (bottom).

Farmers and urban dwellers

Traditionally, farmers have received top priority with water allocation as high as 80 per cent of the available water. Farmers have also been favoured with very low-cost water (often a twentieth of the price of water to industry or households). There is also doubt over the efficiency of irrigation methods and considerable waste. California has upset other states by its aggressive demands on water supply: a lengthy drought plus urban growth has put pressure on other states. Arizona needs more water for the cities of Phoenix and Tucson. Urban California is squeezing agriculture in the Imperial Valley to supply more water to Los Angeles and San Diego.

Environmentalists versus tourists and recreationalists

These two groups have different objectives. Environmentalists are concerned about managing the wilderness and trying to support the needs of wildlife and the natural beauty of the landscape. However, this area attracts large numbers of tourists to walk in the area and enjoy the fantastic scenery and the lakeside and water of Lake Powell.

Native groups and the US Government

Native Americans in the Colorado Basin have, since the 1880s, been in conflict with the Federal Government over long-forgotten treaties and agreements.

Mexican people and the US Government

The Colorado River no longer reaches the sea. Today, 90 per cent of all available water is removed by the seven US states before it reaches Mexico. The delta is contracting, the wetlands disappearing and the river bed is dry. What was once a productive fishing industry no longer exists. Relationships are strained between the US and Mexican Governments over water allocation.

Looking back over the last 80 plus years, what is your conclusion on the success or otherwise of the Colorado River Basin Management project? Do you think this is an awesome engineering marvel with its flood prevention and water sourcing for industry, farming and people that has been successful? Or an expensive, divisive, inefficient and environmentally damaging project? As a geographer you should now have a view on it.

So what will be the need for river basin management projects in the future? Will it be more grand schemes such as the Three Gorges or smaller, more 'friendly' projects? It is felt that most major dam construction in the future is likely to be in the developing countries.

'Earlier this year [2014], the water in Nevada's Lake Mead National Recreation Area dropped to levels it hasn't reached since the 1930s, when the lake was created by the construction of the Hoover Dam.

The lake's slowly shrinking pool due to a combination of a 14-year-long drought in the south-western USA and a dwindling supply of water from the Colorado River puts the nation's largest man-made reservoir at only 39 per cent capacity, as water levels have fallen about 40 m since Mead last reached its peak in 2000.

The US Bureau of Reclamation regional chief Terry Fulp said "water obligations will be met next year without a shortage declaration". The result will be full water deliveries to cities, states, farms and native Indian tribes in an area that's home to some 40 million people including the cities of Las Vegas, Phoenix and Los Angeles. The dropping level since the reservoir was last full in 1998, at just under 400 m above sea level, has left as much as 40 m of distinctive white mineral "bathtub ring" on hard rock surfaces surrounding the lake.

Boaters and swimmers have largely ignored the dropping water levels in a place where splashing in cold fresh water on 40+ °C summer days is a treat. But they've also dealt with marina closures in recent years. Visitors who used to feed scraps to carp from restaurant deck tables may now need to trek hundreds of metres with sandwiches and beach blankets to enjoy a waterside lunch.

Fulp compares controlled management of the two largest reservoirs on the Colorado River to pouring tea from one cup to another. Seven south-western US states are left with the result from a 1928 allocation agreement that also provides shares of Colorado River water to Native American tribes and Mexico. Las Vegas, with more than 2 million residents and about 40 million tourists a year, is almost completely dependent on Lake Mead for drinking water.

Federal and state water officials have negotiated plans for a shortage declaration triggering delivery cuts to Nevada and Arizona. California, Colorado, Utah, New Mexico and Wyoming wouldn't see direct cuts in their share of river water, but officials have acknowledged there would be ripple effects.'

Source Adapted from various website articles, including Weather Underground and Al Jazeera America

Final facts about the Hoover Dam

'Hoover Dam is as tall as a 60-storey building. It was the highest dam in the world when it was completed in 1935. Its base is as thick as two football fields are long. Each spillway, designed to let flood waters pass without harming the dam itself, can handle the volume of water that flows over Niagara Falls. The amount of concrete used in building it was enough to pave a road stretching from San Francisco to New York City. The dam had to be big. It held back what was then, and still is, the largest man-made lake in the USA. The amount of water in the lake, when full, could cover the whole state of Connecticut 3 metres deep. Only a huge dam could stand up to the pressure of so much water.'

Source The National Park Service

▲ **Figure 1.29** Lake Powell showing low reservoir level

 Task

1 As background research, what are the physical characteristics of the Colorado River Basin?
2 Explain the human and physical factors that had to be considered when selecting the site of the Hoover Dam and reservoir.
3 Explain the need for water management on the Colorado River Basin and surrounding areas such as Los Angeles.
4 Describe and explain the social, economic, environmental and political benefits and adverse consequences of the programme. Use Table 1.7 to help you.

▼ **Table 1.7** Summary of the Colorado River Management Basin and the Hoover Dam

Benefits/Advantages	Drawbacks/Disadvantages
Social	**Social**
• A more reliable fresh water supply for the growing desert cities such as Phoenix and Las Vegas • Enhanced quality of life with buildings and shopping malls, all with air conditioning • Many people have swimming pools and landscaped gardens due to the availability of water • Areas such as around Lake Mead provide opportunities for water sport and tourism • Flood control is possible and the threat of loss of life reduced • Jobs have been attracted to the area due to the cheap power from hydroelectric power • Improvements in communications • Food is relatively cheap and plentiful	• Some people were displaced as the dam and reservoirs were filled, although not that many (estimated at 3000 people) • There was a loss of burial sites and other Native American sacred sites following the construction of the dams and the flooding of the valley • As a result of the increasing population, the demand for space, energy and water will continue to grow • Flooding has been reduced. However, in the exceptionally wet 'El Niño' floods of 1983, the dams in the Colorado River Basin were unable to prevent devastation • Increased soil erosion downstream from the dam
Economic	**Economic**
• Agricultural output has increased due to the use of irrigated water. The Imperial Valley, known as the 'salad bowl', allows food surpluses, double cropping and increasing yields • High-tech industries have been attracted to areas such as Phoenix, bringing high-paid employment. Also industries such as aluminium are attracted by the cheap HEP • Farming tends to be very profitable • The urban and industrial growth has meant a boom in the construction industry • Tourism is a massive earner, with both Las Vegas and the Hoover Dam attracting millions of people a year • Fresh water supply all year round leads to increased food output and supports population growth in urban areas such as Los Angeles	• Silting of the reservoirs reduces capacity and dredging is a costly maintenance factor • Loss of silt downstream means that expensive fertilisers are required to maintain yields
Environmental	**Environmental**
• New habitats for birds and wildlife have been created • HEP is considered to be clean electricity and should not add to global climate change • Water can be transferred from areas of surplus to areas of deficit by means of canals, rivers or aqueducts	• High evaporation leads to increased salinity. This shortens not only the life of the turbines but, if left unchecked, will severely impact the soil, reducing crop yield • Salinity is especially an issue as the water moves downstream • The dams trap sediment, which reduces reservoir capacity and must be dredged • Desalination plants have had to be constructed at significant cost ($1 billion for the most recently constructed plant) • There are reports that Rainbow Bridge, a noted natural geological site, is being slowly destroyed by erosion • Significant loss of water from evaporation and seepage • As a result of increased growth there has been an increase in the use of groundwater supplies; this is not sustainable • Plants, birds and animals lost their habitats following flooding of the lakes. The Colorado delta is now a lifeless salt marsh abandoned by wildlife

▼ **Table 1.7** *(Continued)* Summary of the Colorado River Management Basin and the Hoover Dam

Benefits/Advantages	Drawbacks/Disadvantages
Environmental	Environmental
	• The lower reaches of the river no longer carry any fresh water through Mexico to the Gulf of California and the Mexican farmers have problems with irrigation due to the high water salinity • Pollution often moves downstream across borders • Changes in micro-climates caused by the large reservoirs
Political	Political
• There is an agreement (the Colorado River Compact) between the seven US states and the two countries, the USA and Mexico, about the allocation of rights to the river's water • It shows that countries and states can reach agreement on difficult issues	• There have been continuing disputes over the water allocations initially established under the 1922 Colorado River Compact. Water allocations, which affect seven US states and two countries, were often based on unreliable flow rates • Compact agreements concentrate on the quantity not the quality of water

Research opportunity

The area around the Aral Sea has never been a rich, developed agricultural area. However, this basin bears significant environmental problems. Many poor practices, including the diversion of water, farming and uncontrolled industrial development, have resulted in a disappearing sea (now 10 per cent of the area in 1977), salinisation and pollution from farms and factories. These problems, originally an internal issue of the Soviet Union, became a global problem in 1991. Kazakhstan, Kyrgyzstan, Tajikistan, Turkmenistan and Uzbekistan (which all became independent states after the break-up of the Soviet Union) have been struggling since to stabilise and eventually to redevelop parts of the basin.

Until the 1960s, the Aral Sea was the fourth largest inland body of water in the world. Its basin covers 1.9 million km², mainly in what is now Kazakhstan, Kyrgyzstan, Tajikistan, Turkmenistan and Uzbekistan. Parts of the upper basin can also be found in China, Iran and Afghanistan. Two of the major river sources of the Aral Sea, the Amu Darya and the Syr Darya, come from the glacial meltwaters from the high mountain ranges of the Pamir and Tien Shan.

For a thousand years the fertile lands between the Amu Darya and the Syr Darya rivers were supported by irrigation water and channels. Historical records show that the level of the Aral Sea was constant. The sea once had a prosperous fishing industry giving employment to over 75,000 people. However, the Soviet Union planned to change an unproductive part of the region into a fruit and cotton producing belt. Many irrigation projects were started, doubling the area of irrigated land between 1960 and 1990. But there were problems. Such intensive reliance on the cotton crop resulted in environmental degradation. Pesticide use, salinisation and industrial pollution have degraded the quality of the water, giving rise to increased rates of disease and infant mortality (with 10 per cent of children dying in their first year as a result of kidney and heart failure). The Amu Darya and the Syr Darya rivers were reduced to a trickle as the water was diverted to new cultivated irrigated land. Since 1960 the Aral Sea has lost three-quarters of its volume and half its surface area; meanwhile, its salinity has tripled. Salts and chemical residue line the bare seabed, which the winds carry as far as the Atlantic and Pacific Oceans. Over-irrigation has also brought problems with an increased water table level, making drinking water too salty to drink and too polluted to spray on the crops. Wildlife diversity has reduced. The shrinking of the sea combined with local climate change has resulted in the region being affected by greater extremes of temperature and precipitation.

There have been attempts to co-ordinate solutions to the problem. But we have a complex situation involving several countries, changing and dwindling resources resulting in environmental destruction.

An update: 'The Aral Sea is bringing new wealth to fishing villages in Kazakhstan, but their neighbours on the opposite shore in Uzbekistan are suffering a very different fate.'

Dene-Hern Chen , freelance journalist, 23 July 2018

Using the above article, your research task is to update any physical, economic, social and environmental issues impacting this Aral Sea study.

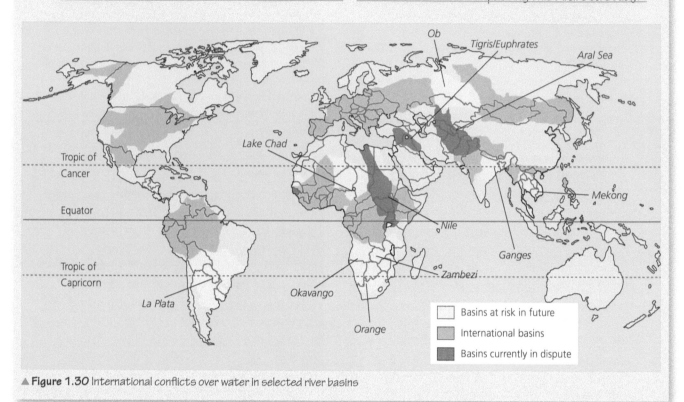

▲ **Figure 1.30** International conflicts over water in selected river basins

Task

1 With reference to Figure 1.12 which shows the major river basins of Africa (page 8), describe and explain the general distribution of river basins across the continent of Africa.

2 Referring to the case study on the Tarbela Dam (page 33):
   a) Explain why there is a continuing need for river basin management.
   b) Explain the physical and human factors which should have been considered when selecting a site for the Tarbela Dam and reservoir.

3 Describe and explain how changes in the storage of water have affected the hydrological cycle of the basin.

Information on 2010 flooding of the Indus Valley:

- Worst flooding since 1994
- Exceptional amount of rain (274 mm in 24 hours)
- 20 million people displaced
- 4 million affected by food shortages
- Bridges over Karakoram Highway destroyed
- Increase in food prices
- Local rioting
- Political unrest
- 2000 deaths

Case study: the Tarbela Dam

The Tarbela Dam in Pakistan is part of the Indus Basin Project (and part of the 1960 Treaty between India and Pakistan). In a question paper case study, you may see unfamiliar maps, figures, tables or charts.

This is not a problem, since you have all the skills and knowledge to answer any question about the physical characteristics of river basins, water management, sites and consequences.

▲ **Figure 1.31** Map of the Indus Valley and Basin, Pakistan

▲ **Figure 1.32** Flood victims, Indus River, 2010

Population

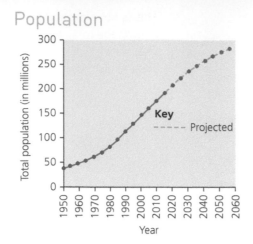

Climate

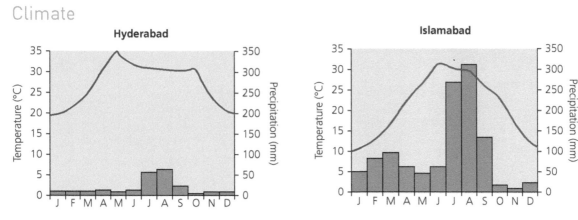

▲ **Figure 1.33** Population and climate information, Pakistan

Summary

In this chapter we have provided you with necessary background information on aspects of the hydrosphere, some within the context of case studies, and a detailed consideration of:

- river basins
- the need for water management
- selection and development of sites
- the consequences of water control projects.

Introduction to Development and Health

In this chapter we will cover two separate but linked topics, those of development and health.

The SQA course assessment specification lists the following topics to be covered in Development and Health:

- validity of development indicators (social, political, economic and composite indicators)
- differences in levels of development between developing countries
- a water-related disease: causes, impact, management
- primary healthcare strategies.

In the first part of this chapter we will consider Development Geography and how we define and measure development. In the second part of the chapter we will examine how levels of health and the incidence of disease are major indicators of development (IoD).

2.1 Introduction to development

'For most people living in Scotland and the UK, life is better today than it was for our grandparents.'

Reflection

What do you think is the evidence for this statement?

Development is about change over time and improvements in the quality of life. A child will develop physically and mature over time. In Geography we concern ourselves with improvement in issues related to society, industry, lifestyle, standard of living, health, education, social justice and happiness. A country which is fully developed can allow its people to reach their potential and to have choices and opportunities.

However, people in economically developing countries do not have the same choices and opportunities. Of course, if you think about any country it is clear that not all citizens are equal. In the UK, wealth, health and the standard of living vary. There are people who have more money than others. It is the same in a less economically developed country (LEDC). For example, in some areas of Marrakech or São Paulo, there are magnificent houses with landscaped gardens and fountains, but within a kilometre you may find some remaining shanty dwellings not even connected to a water supply. Inequality of social and economic development is a major feature of the contemporary world.

Sustainability

In recent years, sustainability has become an important word for scientists, geographers and even politicians. You will encounter several definitions but in Geography we consider sustainability to refer to the needs of future generations and our responsibility not to spoil our landscape, land resources, water resources, atmosphere and oceans. Indeed, we want to live on a sustainable planet.

In Scotland we are privileged to live in a wealthy society where we can be generous and consider the needs of others. This is not a choice that some economically developing countries have. When the UK was going through the Industrial Revolution in the nineteenth century, we valued growth, growth and more growth!

A few inspired people, such as Robert Owen at New Lanark, remembered the health and wellbeing of the workers and the environment. Now that we have moved on, we realise the value of sustainability and care of the environment. Much has been written about the economic growth in countries such as Brazil, India and China. So should we be amazed by the fantastic pace of economic change or dismayed by the stresses on the environment and people?

Task

▲ **Figure 2.1** Inequalities in São Paolo

What does this image show about social, economic and environmental inequalities in São Paulo?

Reflection

Which of these do you think is more important — economic growth, improved quality of life or environmental concern? There are competing views to consider:

- Should we educate the developing nations on how to provide opportunities for their people?
- Should we assist them financially and give greater consideration to their needs?
- Should we assume that, in time, they will be able to use their growth for the improvement of the environment?

As a geographer you have (or will have) the knowledge, skills and energy to make a difference.

Development groups and definitions

There are many ways to classify how developed a country is. Below is a summary of some of the best-known classifications over the last 40 years, but it is worth mentioning that there is no overall agreement about these classifications. By the time you read this book, it may well have changed again!

In the 1970s, countries were classified into the First, Second and Third Worlds. The **First World** included the developed capitalist countries such as the USA, the UK and western European countries. The **Second World** included the communist countries such as the former Soviet Union and the satellite eastern European countries. The Third World included every other country, although in reality this term applied to the group of countries of the world that was recognised as deprived, underdeveloped and not part of the affluent, developed world.

Obviously this was a very basic split and was devised by the Western world. As a model it was too simple and was not useful when it came to describing and even predicting change.

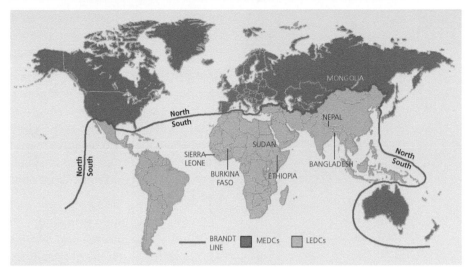

▲ **Figure 2.2** The Brandt line showing the North–South Divide

In 1980, the former West German Chancellor, Willy Brandt, headed the Independent Commission on International Development Issues. Its report, a book called *North–South: A Program for Survival* (widely referred to as the **Brandt Report**), outlined how developed and developing world countries must work together to tackle the problems that faced the world; the term **North–South Divide** was used.

However, Brandt's terms were also too neat and tidy; once again they were deemed to be too simplistic and did not allow for change. Many countries simply did not fit this Brandt model.

So far, neither model could indicate which countries were developing quickly and which were falling further behind.

Case study: the Brandt Report

North–South: A Program for Survival outlined how developed and developing countries must work together to deal with the problems that faced the world. The report forecast that if nothing were done to reduce levels of inequality, then there would be massive problems on a global scale with unemployment, wars, famines and environmental decline. Brandt's group believed that the developing and developed countries relied on each other, and so it was in their shared interests to work together.

The Brandt Report made ten proposals:

1   Developed countries should give technological help to developing countries to allow them to process their raw materials before exporting them, thus earning far more money for themselves.

2   Developed countries should actively encourage industrial development in the developing world countries.

3   Rules should be drawn up to limit the influence of **transnational corporations (TNCs)** in developing countries.

4   A commitment to global disarmament should be enforced.

5   Increased funding should go into the development of new agricultural technology to increase food stocks.

6   Research for alternative energy sources should be increased, as well as increasing the use of HEP. Coal, and especially oil, should be used more efficiently to conserve global stocks. The price of oil should be better controlled and more predictable.

7   More money should be made available by the developed countries for the World Bank to lend to developing countries.

8   A global tax system could help to raise more money for development programmes. The amount of tax paid by a country would relate to its overall wealth.

9   All developed countries should be giving 0.7 per cent of their GNP to international aid by 1985. In reality, only a very few have actually reached this mark (the UK, Luxembourg, Norway, Denmark, Sweden and the Netherlands). The aid should not be conditional or tied in any way.

10 An organisation called the World Development Fund should be set up to co-ordinate the development programmes.

Reflection

Take time to consider whether the Brandt Report's ten-point set of proposals, given on page 37, are relevant today.

Task

'Development will never be and can never be defined to universal satisfaction. It refers to desirable economic and social progress and people will always have different views about what is desirable … economic growth and industrialisation are essential. But if there is no attention to the quality of growth and to the social change one cannot speak of development.'

Source Brandt Report (1980)

'A country, or a village or a community cannot be developed, it can only develop itself. Real development means the development of people. Every country in Africa can show examples of modern facilities which have been provided for the people and which are now rotting unused.'

Source Julius Nyerere (1973)

1   Define 'development' and 'sustainability'.
2   What do you see as the key features of the Brandt Report?

Current classifications of development

Today, countries are classified into more categories to show the continuous spectrum of economic development between the least developed countries and the most developed countries.

By the end of the 1980s countries were reclassified. The idea of capitalism and communism was dropped as a guide to describing development and the key was now economic development. Affluent countries were classified as more economically developed countries (MEDCs) (also known as the rich industrial countries (RICs)) and poorer countries were classified as less economically developed countries (LEDCs).

What do you think about these classifications? The advantage of these terms is that there is a simple split. Some countries are clearly more developed (France and Canada) and some are less developed (Chad and Laos) – no disagreement here. The problem is that for some countries it is difficult to place them with certainty. What about Turkey, Cuba or even China?

When this information was mapped it became clear that there was a divide along North–South lines. The MEDCs were seen to be generally in the North, as they included North America and Europe, but also the southern hemisphere nations of Australia and New Zealand. LEDCs were generally found in the South. Confusing? There's more!

Those early attempts at grouping countries together into categories were less than perfect. So what are we left with today? Over the last 25 years new groups have been recognised:

- **Rich industrial countries (RICs)**: a group of the most developed countries of the world, in some ways similar to the 'North' countries.
- **Newly industrialising countries (NICs)**: a growing group of rapidly developing and getting richer countries, often with an economy based on primary industry (usually agriculture) and a growing secondary industrial base. Examples include China, India, Brazil and South Africa.
- **Least developed countries (LDCs)**: the group of the most economically deprived, least developed countries in the world. The LDCs are not only poor, but struggling to gain any ground on the rest of the world. The list includes Angola, Ethiopia, Malawi and Nepal.
- **Oil-exporting countries (OECs)**: this group includes Qatar, Saudi Arabia and the United Arab Emirates (UAE). These countries have considerable wealth when measured in output and value of exports. The problem is that often this wealth is controlled by relatively few people and the benefits are not shared by all citizens.
- **Former communist countries**: this group includes Poland, the Czech Republic and Romania. These countries are developing fairly rapidly, sometimes as part of the European Union.

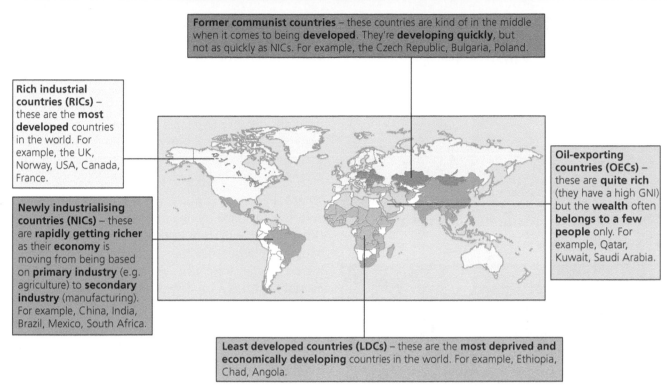

Former communist countries – these countries are kind of in the middle when it comes to being **developed**. They're **developing quickly**, but not as quickly as NICs. For example, the Czech Republic, Bulgaria, Poland.

Rich industrial countries (RICs) – these are the **most developed** countries in the world. For example, the UK, Norway, USA, Canada, France.

Newly industrialising countries (NICs) – these are **rapidly getting richer** as their **economy** is moving from being based on **primary industry** (e.g. agriculture) to **secondary industry** (manufacturing). For example, China, India, Brazil, Mexico, South Africa.

Oil-exporting countries (OECs) – these are **quite rich** (they have a high GNI) but the **wealth** often **belongs to a few people** only. For example, Qatar, Kuwait, Saudi Arabia.

Least developed countries (LDCs) – these are the **most deprived and economically developing** countries in the world. For example, Ethiopia, Chad, Angola.

▲ **Figure 2.3** The development continuum

- **Tiger economies**: this group of rapidly developing Asian economies from the 1980s includes South Korea, Singapore, Taiwan and Hong Kong.
- **BRIC countries**: this is the acronym for a group of four major emerging countries: **B**razil, **R**ussia, **I**ndia and **C**hina. Add South Africa and you now have the BRICS. It is predicted that the BRICS, in economic terms, will overtake the USA, Japan and the European Union by 2050. A prominent bank predicts that China, India, Brazil and Russia will be first, third, fifth and sixth in terms of global economic size within 25 years.

As you can see, there is an almost endless combination of groups of countries. In fact, another new group has recently emerged in the form of **M**exico, **I**ndonesia, **N**igeria and **T**urkey, known collectively as the MINT economies.

2.2 Indicators of development

Using indicators of development (IoD), it is possible to rank every country in the world (approximately 200). A simple comparison would be to a football league table and number of points won. To rank a country we can use a range of indicators:

- economic indicators (wealth, income and jobs)
- social indicators (quality of life, health, education and basic infrastructure)
- demographic indicators (**life expectancy**, population and migration)
- political indicators (democracy/dictatorships, justice and freedom)
- cultural indicators (for example, the role of women, religion).

It is important to note that the indicators of development are changing all the time. Although we have tried to obtain the most accurate and up-to-date set of figures for this book, it is not always easy for a number of reasons. Wealthy countries such as the UK and the USA have a system of collecting information which is generally accurate. As you will find out in your studies of population, for example, collecting information in a economically developing country is important but sometimes not so accurate, and the measures between developed and developing countries vary widely (see the case study on the Human Development Report (HDR) on page 44).

All indicators need to be stated in numerical terms to allow comparison between countries, within countries and over time. So we talk about rates (for example, mortality, birth and death rates), hard numbers (for example, number of years at school, calorific intake, life expectancy), percentages (for example, number of people connected to a sewerage system, adult literacy) and per capita (statistics that are calculated to give a number per person, for example – see Figure 2.6).

What is important is that the statistics can be easily determined and fairly unambiguous in meaning. Often, single indicators can be used as a measure of something larger and more significant. For example, life expectancy from birth gives an indication of the overall health of a country, in other words diet and access to clean water and sanitation.

However, indicators of development such as sustainability, social justice and freedom of speech are more difficult to define and judge and it is impossible to achieve consensus.

There are many different indicators but what is important is that a single indicator should never be used to determine development. Why? Well, indicators usually cover the whole country and there will always be variables within a country, such as between rural and urban areas, or between men and women.

▲ **Figure 2.4** Modern school, Scotland

▲ **Figure 2.5** Overcrowded primary school in Egypt

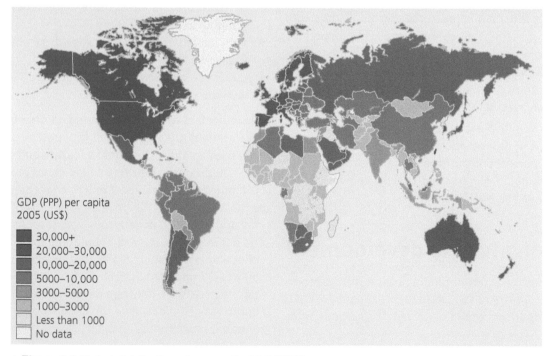

GDP (PPP) per capita 2005 (US$)

- 30,000+
- 20,000–30,000
- 10,000–20,000
- 5000–10,000
- 3000–5000
- 1000–3000
- Less than 1000
- No data

▲ **Figure 2.6** Global distribution of per capita GDP (US$)

Also, as will be shown below, a country might show a very positive indicator in one area but score poorly elsewhere. For example, in Cuba, the number of doctors per 1000 of the population is very high, far higher than in the UK, but it is obviously a less economically developed country when we look at GDP and overall economic indicators.

It is important to give development indicators both their full title and unit of measurement. For example, infant mortality rate is the number of infants who die before they are one year old, per 1000 live births in a given year.

The problem with any development indicator is that it is an *average* figure for the whole country and often hides huge differences in the standard of living within the country.

The most commonly used measures of development are gross domestic product (GDP) and gross national income (GNI). These are both usually expressed per capita, meaning the total is divided by the number of people living in the country to obtain a value per person.

- GDP is the total value of all finished goods and services produced by a country in a year.
- GNI is the total value of all finished goods and services produced by a country in a year, plus the net income earned by that country and its population, including overseas investments.

Both GDP and GNI are usually expressed in US$ so that comparisons can be made.

Levels of development

In general, reasons for the differences in levels of development relate to both positive and negative factors and to both physical and human factors.

- Positive factors include a good location for trade, a mild climate or attractive scenery to encourage tourism, natural resources like oil or copper, a stable government and an educated workforce.
- Negative factors often include remoteness, an extreme climate (cold/hot/dry/wet), disease, an absence of natural resources, a corrupt or dictatorial government, civil war, being prone to natural disasters such as drought, cyclones, floods or earthquakes, a fast-growing population and debt.

In the exam you may be asked to compare development only within the developing world. We have included data from the developed world since it can be useful in understanding differences.

 Task

▼ **Table 2.1** Comparison of indicators of development for Malaysia, Mali and the UK, 2018

Development indicator	Malaysia	Mali	UK
GDP per capita (US$)	11,500	762	42,000
Doctors per 100,000 people	150	14	280
Adult literacy (%)	93	33	99
Birth rate per 1000 people	17	42	12
Life expectancy (years)	75	58	82

Table 2.1 compares a number of development indicators for Malaysia, Mali and the UK. Look closely at the values for each development indicator. What does it tell you about life in that country? What explains the huge differences between these three countries?

Within the developing world, different countries are at very different stages of development. Information about countries' development is often presented in a table showing different indicators of development. Table 2.1 suggests that the three countries shown are at different levels of development. It does not matter if you do not know anything about the UK, Mali or Malaysia – what is important is an understanding of what the indicators represent and an ability to recognise differences.

- GDP per capita (US$) is an economic indicator showing the wealth of a country, which in turn shows level of development. The figures in Table 2.1 show that Malaysia has more than 15 times the wealth of Mali but far less than the UK.
- Doctors per 100,000 people is a social indicator of health provision in a country. The table tells us that Malaysia has around half the number of doctors per 100,000 people as the UK, whereas Mali has very few doctors per 100,000 people – only 14!

- Birth rate per 1000 people is a social indicator which can also show population trends and therefore level of development. The birth rate in a country falls as it develops. The figures show the UK as the most developed and Mali the least developed.

- Adult literacy is a social indicator showing level of education. Mali has a very low level of literacy compared with the UK and Malaysia.

In conclusion, the data in the table show that these three countries are at very different levels of development.

 Task

1 Infant mortality rate per 1000 live births is a social indicator of development. Name two other social indicators and two economic indicators and explain how they show a country's level of development.

2 a) 'The percentage of households with a connection to the internet is an example of a social indicator.' Explain what this means. In your view, is this a good indicator of social development? Justify your answer.

   b) Suggest reasons why there is such a wide range in levels of development as measured by this indicator.

   c) This single indicator is an indication of more than internet access and use. Suggest how this information could be linked to wider features of development and standard of living.

3 a) Using Table 2.2, describe and explain the variations between the UK and Afghanistan.

   b) Within the developed world, why do you think there is such a variation between Iceland and the USA?

   c) Suggest reasons for the variations between Japan, Russia, Bangladesh and India. How do you think the percentage of internet users will change over the next 25 years?

▼ **Table 2.2** Global internet use 2018: percentage of households in selected countries with an internet connection

Country	% internet users	Country	% internet users
Iceland	99	Russia	76
Sweden	97	Romania	75
UK	94	China	54
Japan	93	Bangladesh	48
New Zealand	89	India	34
USA	88	Ethiopia	15
Portugal	78	Afghanistan	15

Source www.internetworldstats.com

Composite indicators

It is widely believed that individual indicators are of limited value, since they tend to be 'average' figures and conceal large variations that exist between and within countries. A country with an above average calorific intake indicator may have well-fed people, but this does not tell us much about their overall health, education or social freedom. For that reason it is necessary to group together indicators.

The UN argues that development is about improving people's social as well as economic wellbeing. It is possible that people may have low incomes yet have a decent quality of life if they have access to free education, clean water and free healthcare. Every measure of development has merits and limitations. No single measure provides a complete summary of the differences in development between countries. Composite indicators have two or more separate indicators, resulting in a fuller and more

representative description of development in a country.

Composite indicator 1: Physical Quality of Life Index (PQLI)

M.D. Morris is generally accepted as the creator of the Physical Quality of Life Index (PQLI) in the mid-1970s. This composite indicator summarises infant mortality, life expectancy at age one and basic literacy on a scale from 0 to 100. Countries are ranked, not by income, but by real changes in real-life chances.

Infant mortality, life expectancy at age one and basic literacy are central to the wellbeing of the most economically deprived. These indicators can also be used as substitutes for linked concepts for development. For example, if money is invested in industrial, educational and health development, then that will be reflected in improvements to the PQLI.

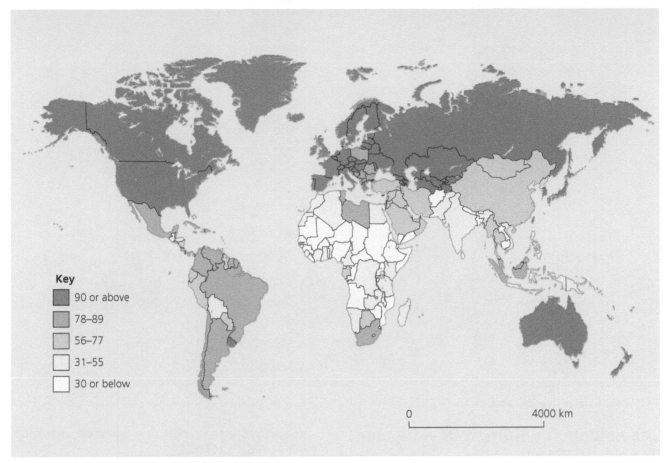

▲ **Figure 2.7** Physical Quality of Life Index (PQLI): a global perspective

Key

- 90 or above
- 78–89
- 56–77
- 31–55
- 30 or below

0   4000 km

However, there are criticisms of the PQLI. There is an arbitrary nature to the mathematical formula used to convert percentages and raw numbers to a scale. Also, the PQLI ignores economic factors and other 'quality of life' areas such as justice, security and human rights. It has also been suggested that there is overlap between what infant mortality and life expectancy measure.

The PQLI, however, is generally accepted as a realistic and positive measure of change and success. It can be used to compare countries at a fixed time or changes in a country over time. A figure of around 80 is now considered to be satisfactory. In 1960, 53 per cent of the world's population lived in countries with a PQLI of less than 50. By 2000 it was down to 10 per cent and latest figures calculate that the figure is now around 8 per cent. Although it demonstrates that positive progress is being made, it also now reveals a small group of economically deprived countries.

Composite indicator 2: Human Development Index (HDI)

The Human Development Index (HDI) uses an adjusted income per capita (income with consideration to purchasing power), educational attainment (combination of adult literacy rates and average number of years of schooling) and life expectancy at birth. The scale is from 0 (worst) to 1 (best). The HDI was devised by the UN to describe human development (both economic and social) within and between countries.

Every year the UN publishes the Human Development Report (HDR), which uses HDI to rank all the countries of the world in their level of development. Since 1990 this report also comments on other global challenges such as gender, poverty, climate change, globalisation, democracy and human rights.

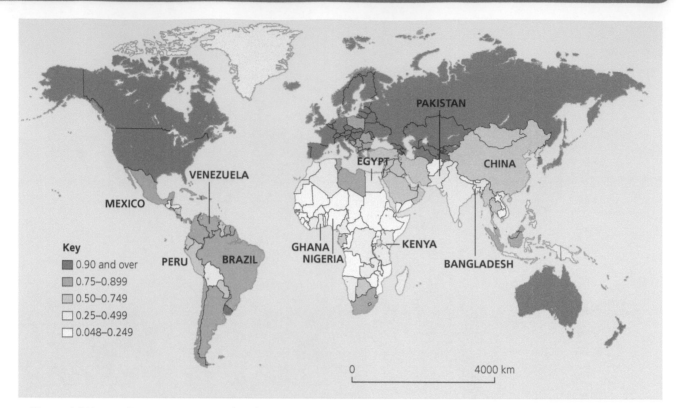

▲ **Figure 2.8** Human Development Index (HDI): a global perspective

Case study: the Human Development Report (HDR)

The HDR is a weighty and important set of data. It is the generally accepted core report charting change and progress and covers many themes and issues:

- **Life expectancy**: the average age to which a person lives, for example this is 82 in the UK and 67 in Kenya.
- **Infant mortality rate**: the number of babies, per 1000 live births, who die below the age of one. This is 4 in the UK and 36 in Kenya.
- **Poverty**: indices count the percentage of people living below the poverty level or on very low incomes (for example, less than $2 per day).
- **Access to basic services**: the availability of services necessary for a healthy life, such as clean water and sanitation.
- **Access to healthcare**: takes into account statistics such as how many doctors there are for every patient.
- **Risk of disease**: calculates the percentage of people with diseases such as HIV/AIDS, malaria and tuberculosis.
- **Access to education**: measures the percentage of people who attend primary school, secondary

school and higher education; this is often linked with gender statistics.

- **Literacy rate**: the percentage of adults who can read and write. This is 99 per cent in the UK, 79 per cent in Kenya and 69 per cent in India.
- **Access to technology**: includes statistics such as the percentage of people with access to phones, mobile phones, television and the internet.
- **Male/female equality**: compares statistics such as literacy rates and employment between the sexes.
- **Government spending priorities**: compares health and education expenditure with military expenditure and repayment of debts.

In the West, we can be very negative about change and development on our planet. However, we need to be aware that recent changes have been essentially positive, although there is still a lot of work to be done.

A summary of recent HDI findings indicates that poverty has been reduced in some respects in almost all countries.

- Poverty has fallen more in the last 25 years than in the previous 200 years. When comparing countries, by looking at GDP for example, the gap has widened between the economically developing countries more than the affluent countries. More nations are falling behind than catching up. However, countries such as India and China have made immense reductions in poverty.

- Figures also compare development *within* a country. By looking at these figures we can see that the variations within countries such as China or Brazil can be massive. The gap between the richest and poorest people in China is widening. What is the relationship between poverty and inequality?

- How can we have an overall improvement in levels of development and at the same time a widening gap? Possibly the answer is that there appears to be a core resistant group unable to make any progress in the improvement in their quality of life.

- However, two issues are persistent in giving us cause for concern. First, approximately one person in six worldwide struggles on a daily basis in terms of sufficient nutrition, clean and safe drinking water, adequate shelter, basic sanitation and access to healthcare. Second, there is an increasing level of disadvantage, known as the **development gap**.

▼ **Table 2.3** Level of human development: HDI changes, 1990–2018

Level of human development	HDI value (0 to 1)	1990 Number of countries	2010 Number of countries	2018 Number of countries
Very high	0.900 and over	8	10	17
High	0.800 to 0.89	23	29	42
Medium	0.500 to 0.799	82	101	99
Low	Below 0.500	73	46	28

Composite indicator 3: International Human Suffering Index (IHSI)

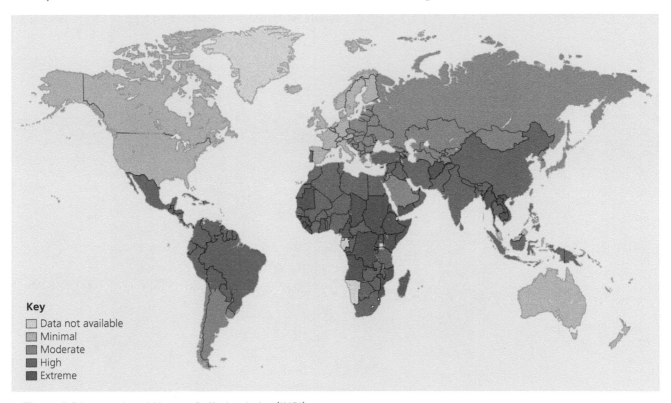

Key
- Data not available
- Minimal
- Moderate
- High
- Extreme

▲ **Figure 2.9** International Human Suffering Index (IHSI)

Research opportunity

Conduct your own research into the International Human Suffering Index. The Population Crisis Committee in Washington DC suggests that the index shown in Figure 2.9 is a compound indicator of distress, compiled by 'adding together ten measures of human welfare related to economics, demography, health and governance: income, inflation, demand for new jobs, urban pressure, infant mortality, nutrition, access to clean water, energy use, adult literacy, and personal freedom.'

Task

What are the main elements of the PQLI composite indicator of development? Explain why such composite indicators are more useful in establishing a country's level of development than single indicators.

It is possible to describe dozens of recognised composite indicators. What you need to know is what they are, why they add more to the study than single indicators, some key examples, and their strengths and weaknesses. A final note on indicators: we even have an index on 'happiness' – this can be researched online. The UK Government is keen to investigate this feature, to compare happiness across countries of the world and changes over time.

2.3 Development gap

The gap between the level of development of the world's wealthiest and poorest nations is widening. What is the evidence for saying this? This gap, once identified as the North–South Divide, can be recognised using indices such as the HDI and PQLI, but it can also be seen *within* countries in terms of the urban–rural split and through studies in the context of gender, ethnicity and religion. It is in all our interests that these gaps are reduced.

The map shown in Figure 2.10 reveals that most of the world's least developed countries (LDCs) are in sub-Saharan Africa, although a few are found in Asia, such as Afghanistan and Nepal. There is a huge

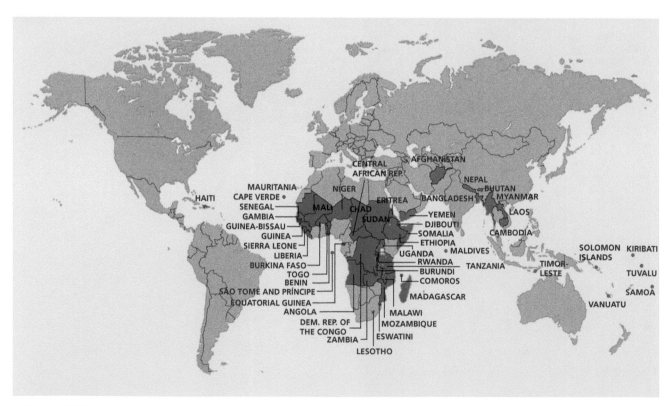

▲ **Figure 2.10** Least developed countries

gap in the stage of development between the most developed countries and the LDCs. Recent research indicates that the wealthiest 20 per cent of countries have almost 85 per cent of the global economy and the most economically deprived 20 per cent have only 1 per cent of the global economy.

The LDCs remain less economically developed because of physical, economic, social, demographic, political and cultural reasons. Clearly, any one country does not have all of these issues. However, these characteristics contribute to a low quality of life for the people living in the LDCs.

Figure 2.11 illustrates when factors combine to produce higher and lower levels of development.

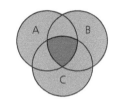

a) Development differences: more advanced developing countries

b) Development differences: least advanced developing countries

■ Highest level of development

■ Lowest level of development

A: Largest countries in region

B: Countries with abundant natural resources

C: Newly industrialised countries

A: Landlocked or island developing countries

B: Countries with few natural resources

C: Countries seriously affected by natural hazards

▲ **Figure 2.11** Factors affecting rates of development

In diagram a), when three factors (size, natural resources and industrialisation) are present there will be fast development. In diagram b), the area where the three factors overlap (positioning of country, few resources and susceptibility to natural hazards) is where least development will be experienced.

Would it not be wonderful to solve this problem? Surely we have the knowledge and desire to reduce the overall level of human suffering? Unfortunately it is not this simple. An economist, Jeffrey Sachs, has argued that there is no 'one solution to fit all struggling nations'. Each nation must have its individual development prescription: there is no single 'cure-all'.

2.4 Differences in levels of development between developing countries

In the SQA Higher Geography examination, if there is a question on differences in levels of development, it will focus on developing countries.

The SQA question will be like this example:

*'Suggest reasons for the wide variations in development which exist **between** developing countries. In your answer refer to countries that you have studied.'*

Your answer should be relevant to the question and should only use *developing* countries. The depth and detail of your answer will be linked to the number of marks awarded, but in general this style of question demands considerable detail and a range of examples.

Factors that influence the rate at which a country may develop can be physical, social, political, economic or environmental. Understanding the reasons why a country is classified as developing is important, as it helps us to understand what factors may help that country to develop further.

Table 2.4 (page 48) and Table 2.5 (page 53) are useful summaries of the factors that influence variation in development between developing countries. The points mentioned in Table 2.4 can be positive or negative. For example, a country which is remote or landlocked is disadvantaged. A country with a coastline or that is close to developed or industrialised countries has an advantage. Note that there is overlap between some of the examples below. For example, the status of women and human rights can be both political and social.

▼ **Table 2.4** Physical, social, political, economic and environmental differences between developing countries

Physical	Factors
Climate	Too hot, too dry, too wet or too cold. For example, the Sahel region of Africa suffers from insufficient rain and high **variability of rainfall**, which means that **drought** is common. Arid areas such as the 'empty quarter' of the Arabian Peninsula offer few opportunities.
	Some **diseases**, such as yellow fever (in Guinea or Uganda) and malaria, thrive in tropical climates due to the hot and humid conditions.
	When climatic elements are positive, development is possible.
Relief, drainage and harshness of the environment	**Mountains** and **steep slopes** discourage development (such as in Bolivia or Bhutan), making it difficult to farm and make a living.
	Areas with **poor drainage**, such as the marshlands of Botswana or the Pantanal (Paraguay), offer limited opportunity for development.
Natural hazards	Within some developing countries there are areas that are prone to **floods** (Bangladesh), **drought** (Sudan), **earthquakes** (Peru), **tsunamis** (Philippines), **volcanoes** (Papua New Guinea), **tropical storms** (Cuba) and **wild fires** (Angola). These hazards can limit future growth and destroy built and agricultural areas. A country may have to divert income to help recovery from natural disasters.
Landlocked, remote or isolated	There are 16 **landlocked** countries in Africa, such as Niger and Zambia (Figure 2.12 on page 50). It is difficult for them to trade, as goods must be driven through other countries to get to the coast for shipping. Countries such as Mongolia and Uzbekistan, while rich in minerals, are **remote** from the main industrial zones of the world.
Mineral resources	The presence of mineral resources can have a significant impact on a country. For example, **oil** and **gas** in Saudi Arabia and the United Arab Emirates, and **copper** in Morocco and the Democratic Republic of the Congo, have brought localised development. Some countries, for example Mali and Laos, have few or undeveloped resources.
	Water is essential for health. Globally, one in six people does not have access to safe water. If water is unsafe, people may be unable to work or care for their families because of illness. Countries with water cleanliness issues include Haiti, Cambodia and Ethiopia.
Human	**Factors**
Social	*'The biggest enemy of health in the developing world is poverty. Education is a human right with immense power to transform. On its foundation rest the cornerstones of freedom, democracy and sustainable human development.'*
	Kofi Annan (former UN secretary-general)
	A developing country finds it more difficult to invest in **education**. The problem is made worse because many countries have a high dependency ratio. Having money to invest in a **healthcare** system is important for a country to develop, because sick people cannot work hard. Within the Caribbean, there are strong contrasts. Haiti has low levels of education and healthcare, whereas Cuba has invested heavily in both of these social indicators.
	The pace of **population growth** and the capability of a country to cope with this pace is another factor. Some countries, such as Ghana, have a population structure with 44% of the population under the age of 15, resulting in a high dependency ratio. Thailand, in contrast, has potentially successful population strategies in place.

▼ **Table 2.4** *(Continued)* Physical, social, political, economic and environmental differences between developing countries

Human	Factors
	High population growth generally limits development as scarce resources must be allocated to a greater number of people, affecting access to water, food, power, schools and health clinics.
	There is often an imbalance within a country between the life chances and opportunities of **men** and **women**, and between people living in **urban** and **rural** communities. Within Tunisia for example, women in rural areas have fewer opportunities to continue in education.
	Ethnic and **religious diversity**, together with any associated disruption, can hold back development. For example, in Tibet there is significant unrest between indigenous Tibetans and migrants from south-west China (supported by the Chinese Government). In Myanmar, there is evidence of ethnic cleansing by the government forces of the Muslim minority, the Rohingya.
	The reduced status of women and a lack of human rights can be observed in several Middle East countries, such as Saudi Arabia and UAE.
Political	A **government** can attract foreign direct investment if it is stable (e.g. Morocco) or discourage investment if it is unstable (e.g. Libya or Iran). Instability can result in **war** and **conflict** (Syria and Yemen). Some nations, such as Nigeria, Columbia and Guatemala, are associated with **corruption** and some are **ethically sound**, such as Bhutan. Money that could be spent on development may be used to fund **military weapons** (for example, Saudi Arabia, Sudan or Algeria) or the **affluent lifestyle** of an elite group of people (for example, the corrupt generals in Myanmar).
	There are many examples of **civil war** in the world today. Since 2010, research has indicated over 40 countries with internal fighting between different groups in countries such as Mozambique, Mali, Sudan, Yemen, Ukraine, Central African Republic, Egypt, Pakistan, Syria, Iraq, Yemen, Myanmar and many more.
	Many people believe that the impact of **colonialism** has hindered development in developing countries. The occupying countries (the colonialists, such as Britain and France) exploited their colonies economically by exporting their food and minerals. There was some investment in the colonies, but this was focused on things that would help trade between the countries. Borders of some colonial countries were set without attention to **tribal** and **cultural differences**, causing tensions and instability (for example, Ghana). (Check www.quora.com for opposing views of colonialism.)
Economic	As Willy Brandt stated in 1980, there is no fairness in global trading (page 37). Developing countries tend to sell **primary produce**. For example, Indonesia and Bolivia trade in tin. In the world trade for rice, farmers compete which lowers the prices they receive. There is more money to be made in processing goods. For example, South Korea trades in electronic goods.
	Foreign direct investment (FDI) can help a country to develop. African countries receive less than 5% of FDI but has 15% of the world's population. European countries receive 45% of FDI and only has 7% of the world's population. Who *controls* world trade is also important – developed countries control the most trade. Morocco benefits from investment from EU countries, which Libya misses out on.
	Many developing countries are in **debt** to developed countries, so some of their income is used to make payments (linked to political issues). Other factors include being members of **trading partnerships** or economic groupings of countries, penetration by **transnational corporations**, influence of globalisation trends and the extent of infrastructure within the country (for example, roads, telephones, Wi-Fi, power).

▼ **Table 2.4** *(Continued)* Physical, social, political, economic and environmental differences between developing countries

Human	Factors
	Another factor that affects development is **employment structure**. Countries with a high percentage of people employed in agriculture tend to be at the lower end of development (for example, Burkina Faso and Nepal).
	Around 40% of the global population depends on agriculture, which makes it the largest employment sector in the world. Malawi, for example, has a high proportion of the workforce in **subsistence farming**, which does not bring income into the country.
	Countries with a poor education system have many **low-skilled workers** and are unable to attract FDI. Countries which have accumulated large debts have to repay **loans** and **interest**, depleting money for services.
	Some countries are over reliant on one or two low-value **exports**. If anything happens to the production or price of these export products, the country's economy is hit badly. For example, Saudi Arabia exports oil which is in high demand, whereas Burkina Faso exports shea nuts which are used in cosmetics.
	Some governments have managed to attract **investment** and increase their manufacturing industries (for example, the Asian Tiger economies of Taiwan and South Korea) through having low labour costs (no trade unions) and lax pollution laws.
	Some countries (for example, Barbados) have **tourism potential** that creates job opportunities and attracts income. South Korea (arguably now developed) has an educated, resourceful and relatively **cheap labour force** that attracts international investors, which in turn allows investment in healthcare, education and infrastructure.
	Countries with a poor **education** system, such as Angola and Mexico, have many low-skilled workers and are unable to attract FDI.
	Lack of government regulation and **lax pollution laws** in parts of south-east Asia have also made some countries (such as parts of urban India) more attractive for manufacturing industries.
Environmental impact	The **ecological** and **social degradation** happening due to the industrialisation of farming includes erosion, rapid deforestation, pollution of soil and water resources, and reduction in biodiversity.
	Some countries have beautiful beaches and wonderful scenery, which attract **tourists** and create job opportunities, such as in Jamaica and Thailand. But there are growing concerns regarding the negative impact of large-scale tourism on the environment, particularly in regions that are rich in biodiversity and where the ecosystem's delicate natural balance is threatened.

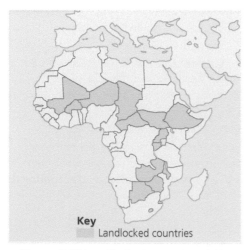

Key
Landlocked countries

▲ **Figure 2.12** Landlocked countries of Africa

Consequences of the development gap

The development gap has consequences for people in the less developed and economically deprived countries. Poverty is the most obvious feature, but how do we define poverty? A basic definition is being unable to have access to the essentials of life, namely water, shelter, sanitation, healthcare and education. Poverty can be measured using indicators of development. In 2014 the World Bank suggested that the poverty line should be set at an income of less than $2 a day (approximately £1.50 at 2018 rates). Hardly an ambitious level! The United Nations (UN) launched the Millennium Development Goals (MDGs) project in 2000 to reduce global poverty by 50 per cent before the end of 2015.

Other goals were to:

- eradicate extreme hunger and poverty
- achieve universal primary education
- ensure environmental sustainability
- reduce child mortality
- promote gender equality and empower women
- combat HIV/AIDS, malaria and other diseases.

The UN set very precise targets. So what is the progress to date? The UN Human Development Report will continue to chart the progress and movement in all these areas, and there have been advancements in levels of overall growth, but two significant events hit hard from 2008 to 2011.

With the collapse of many banks and the crisis that followed, growth in the West was checked and the world went into a downward economic spin. Also around 2006 to 2008–09 there was a global food crisis that saw a steep rise in the price of basic foods in most of the developing nations, with rice, wheat, maize and soya almost doubling in price. We know why this happened:

- drought and climate change resulted in some areas of the world experiencing fluctuating output
- changing and increasing price of oil resulted in increased fertiliser and fuel costs
- loss of agricultural land to the new growth biofuel crops
- dietary changes towards more meat and less grain
- population growth continuing in some of the economically developing areas of the world.

As a result of these factors, the UN became aware that the MDG project may well fall short of its target.

Source Researched & adapted from www.un.org and the UN Millennium Development Goals reports

Reflection

Does poverty exist in developed countries such as Scotland? How can poverty here compare to poverty in Angola when we compare GDP, life expectancy, access to clean water and so on?

In view of this, it is useful to introduce two additional concepts: that of absolute poverty and that of relative poverty. 'Absolute' describes the worst situation in the LDCs. Hunger here is absolute. There is little or no food with no surplus or stored supplies. People will experience uncertainty over where and when their next meal will come from. 'Relative' poverty is when we compare life *within* a country, where levels of equality are uneven. Living standards and expectations are rising all the time in a society such as in the USA or Europe. Some people will share in the wealth of a country, others will not. Every town and city in the UK will have people without a home or a job. In a rich society they experience real, but relative poverty. Would you consider your mobile phone as an absolute basic in life?

Research opportunity

The persistence of poverty is remarkable. Think of all the advancements in the last 30 years: economic growth, increased levels of international aid, debt relief and increased global investment, growth of transnationals, increased global food output, better communications, increased levels of power development and so on.

Why, therefore, does poverty remain a global feature today?

Compare your 'quality of life' today with that of your parents/grandparents when they were your age – ask them about it!

▲ **Figure 2.13** Poverty remains a huge problem in many countries around the world. Here young children from Kibera slum in Nairobi, Kenya, are collecting water for drinking.

Contrast in development and the development gap: a summary

We have identified a range of factors that can influence levels of development both within and between countries. We have also recognised what is referred to as the development gap.

Development continuum

As we have seen, countries are classified into categories that show a continuous spectrum of economic development from the least developed countries to the most developed countries. This has been referred to as the development continuum.

A key feature of the process known as globalisation is the growth and emergence of a group of 'newly industrialised countries' (NICs). This term covers the dynamic growth of a group of countries that, since the early 1970s, has undergone rapid industrialisation and economic expansion. In the early 1970s some transnational corporations (TNCs) were looking to expand and the countries of south-east and east Asia were targeted mainly due to lower costs (especially labour costs). In this first phase of growth, countries such as South Korea, Singapore, Taiwan and Hong Kong (collectively known as the Asian Tigers) were all transformed into modern industrial states.

As geographers, we want to know *why* a situation occurs, as well as its advantages and disadvantages. The Asian Tiger countries tended to have most of the following criteria:

- A well-developed infrastructure of communications (roads, ports and railways), power and water.
- A well-educated, hardworking, low-labour-cost workforce with a range of skills ideally suited for manufacturing industries such as textiles, toys, footwear and clothing.

- A strong, positive, global location suitable for trading (for example, on the Pacific rim with strong links to the Americas and Asia).
- Political stability, allowing investment security, low interest bank rates, a government that broadly favoured Western economic structures and a good chance of profit on investments. FDI was welcomed.
- A relaxed approach towards labour regulation, taxation and (increasingly) lower levels of concern over pollution and the environment, together with a lack of trade union activity and strikes.
- Reduced trade tariffs, making importing and exporting cheaper.

In time these countries developed their own huge conglomerates. Tiger countries became major investors in their own country and sophisticated, high-value new products became a major feature. In the West, traditional industries started to lose out. For example, the UK was hit by competition in the ship-building and vehicle industries, with closures and job losses.

By the 1990s, a second phase of development was well under way. The Tiger economies experienced rising labour and industrial costs and their advantages were not so prominent. The industrial West and even the new TNCs from the Tiger economies began to look for new countries that now had the lowest labour and industrial costs. As a result, countries such as Malaysia, Thailand and the Philippines began to emerge.

By the late 1990s and the early 2000s, the third phase was evident. China and India emerged as the target for FDI and both these countries began and have continued to show impressive industrial growth.

Rapid growth in these countries has led to a number of issues and conflicts. Of course increased development has resulted in increased wealth, reduced levels of poverty and improvements in health, education and welfare. However, there are many reports indicating concern over the exploitation of labour, especially that of children and women. In Malaysia and Singapore there was mass immigration of illegal migrants looking for work, most of whom were employed in the most dangerous jobs. In Borneo, for example, rapid expansion has caused considerable damage to the environment through deforestation and land degradation.

In addition, since these newly industrialised economies tend to export to the West, the economic recession and banking crisis that hit the world from 2008 to 2013 particularly affected some of these economies. Finally, corruption has been a feature of both rich and economically deprived countries.

Table 2.7 (page 55) provides a summary of the different ways of classifying the world into development groups.

Reflection

Transnational corporations (TNCs) are companies that produce, sell or are located in two or more countries and they have an important role to play when we are considering issues of development. Think of Coca-Cola – manufactured in over 100 countries and sold in almost every country in the world, the Coca-Cola logo and brand are recognisable globally. TNCs are found in all categories of industry, including primary (TNCs have links with farming and mining), secondary/manufacturing and in the service/tertiary areas of finance, insurance and banking.

Are TNCs good or bad? It really depends on your own point of view. Think about this carefully; you might find it helpful to discuss this with other students and may even like to debate the issue.

Task

Table 2.5 shows the four key factors that affect development – physical, social, political and economic – as well as a range of issues within each factor. From this it is possible to write up a report on each country of the world to determine their unique point of development. Look at Tables 2.5 and 2.6 and then answer the questions that follow them.

▼ **Table 2.5** Factors affecting differences in levels of development

Factor	Examples
Physical	• Water availability • Agricultural potential • Mineral resources • Harshness of environment • Natural soil quality • Natural disasters • Climate • Geographical isolation • Land degradation • Landlocked location
Social	• Population structure • Dependency levels • Birth rates and population policies • Infrastructure quality • Education levels • Workforce skills • Health of population • Comparison of urban to rural population • Potential for tourism • Caught in 'vicious cycle of poverty'
Political	• Government stability • Level of overseas investment • Level of debt • Corruption • Ethnic and religious tensions • Colonial legacy • Gender inequalities • War and conflict • Political allies • Human rights
Economic	• Stage of industrial development • Dependency on key industries • Globalisation trends • Infrastructure (roads, telephones, internet, power) • Penetration by TNCs • Distribution of wealth • Trading links • Economic groupings • Agriculture: subsistence or cash

▼ **Table 2.6** Selected countries: Human Development Index

World rank	Country	HDI (2018)
1	Norway	0.953
6	Iceland	0.935
9	Singapore	0.932
13	USA	0.924
14	UK	0.922
22	South Korea	0.903
49	Russia	0.816
73	Cuba	0.777
74	Mexico	0.774
79	Brazil	0.759
86	China	0.752
113	South Africa	0.699
130	India	0.640
136	Bangladesh	0.608
142	Kenya	0.590
157	Nigeria	0.532
168	Afghanistan	0.498
182	Mali	0.427

1 Explain the advantages of using composite indicators such as the HDI over single indicators.

2 Select three countries from Table 2.6, one from the top, one from the bottom and one from the middle. Identify some of the factors from Table 2.5 that may explain their levels of development.

▼ **Table 2.7** Summary of the different ways of classifying the world into development groups

North/South Divide	First World to the Fourth World	MEDC or LEDC	LDC	NIC	Tiger economies	BRIC(S) and MINT	Former communist countries
Brandt Report, 1980	First World: richer countries – mainly Europe, North America, Australia Second World: former communist countries such as Russia	More economically developed countries and less economically developed countries	Least developed countries	Newly industrialised countries	A south-east Asian group of rapidly developing economies	BRIC(S): a new grouping reflecting rapid development – Brazil, Russia, India, China, (South Africa)	Usually eastern European, once part of the Soviet Bloc, now independent
The 'Rich North' and the 'Poor South'	Third World: the less economically developed countries of the world. Similar to LEDC and 'The South' Fourth 'World: a new term applied to economically deprived countries such as Chad or the Congo	MEDC linked very closely to the First World and 'The North' LEDC similarly linked to 'The South' and the Third and Fourth Worlds	Similar to the Fourth World and part of the 'South'	This is a growing group of rapidly developing and getting richer countries. For example China, India and Brazil	South Korea, Singapore, Taiwan and Hong Kong	MINT: another new group beginning to emerge – Malaysia, Indonesia, Nigeria and Turkey	Varying levels of development. For example, Poland, Czech Republic and Bulgaria

Theories of development

Geographers have suggested a number of theories explaining why countries are at different stages of development and the widening development gap between the economically developed and developing countries. They concentrate on the ways in which variations in wealth can be found in different economic, social, human and political systems.

In the 1980s, development and industrialisation were considered to go hand in hand. To those in the South, the developed North had a high standard of living and economic growth. Since the North relied heavily on an industrial base for this wealth, it was naturally assumed by both the North and South that the path towards development was one based on an industrial foundation. It was believed that, through industry, the whole economy would be regenerated and that this would result in improvements and efficiencies in agriculture and that higher standards in social conditions would follow.

Figure 2.14 is a model illustrating the possible paths for development. Whatever path is chosen, a source of finance is necessary. No country has to select one path. It is possible to embark on a number of routes to give a broad base to reach the chosen development.

DEVELOPING STATE		DEVELOPED STATE
Route A	Industrial path	
Route B	Agriculture path	
Route C	Development of resources	
Route D	Trading on global scale	
Route E	Overcoming debt	
Route F	Political stability	
Route G	Tourism path	
Route H	Development of infrastructure	
Route I	Health investment path	
Route J	Education path	
Route K	Social/gender equality	
Route L	Employment opportunity	
Route M	Population stability path	

▲ **Figure 2.14** Pathways to development model

Research opportunity

Economic and political powers are not evenly distributed; there is a clear imbalance as a result of complex stories and histories. Power can be shown to work through trade, by military action and use and control of resources. The topic of power is a major global theme worthy of its own chapter, but some key issues can be covered here in this chapter. Power can be exercised at a global level, at national level and at local level. Power patterns are constantly changing; the USA is the clear global superpower at this point in time, although the European Union and China may attain this status in the future. Russia, once part of the USSR, lost its super status but, along with India and Brazil, will continue to gain increased power and influence. The idea of **cultural hegemony** was first developed by an Italian, Antonio Gramsci, 100 years ago. It describes a situation where a government, such as in the USA, can have an influence over the beliefs, values and views of the people in that country. Without the use of force or manipulation, this idea becomes the norm. It means that in such conditions, the ideology of the powerful becomes the accepted and unchallenged way to see the world. US consumerism and cultural hegemony is globally dominant at present.

Table 2.8 shows the changing pattern of the most economically powerful global companies (by $ value). It is not unusual for political activists to protest at major gatherings of the leading nations of the world (for example, the G7 group) and target some of these 'icons' of capitalism (for example, McDonald's). US companies continue to dominate, with Apple and Google replacing Coca-Cola as top brand companies.

▲ **Figure 2.15** Fast food, Indian style!

▼ Table 2.8 Top ten global brand values, 2013 and 2017

	2013	Country of origin	Sector	2017	Country of origin	Sector	Value in 2017 US $m
1	Coca-Cola	USA	Beverages	Apple	USA	Computer services	184,000
2	IBM	USA	Computer services	Google	USA	Internet services	141,000
3	Microsoft	USA	Computer software	Microsoft	USA	Computer software	80,000
4	General Electric	USA	Diversified	Coca-Cola	USA	Beverages	70,000
5	Nokia	Finland	Computer electronics	Amazon	USA	e-commerce	65,000
6	Toyota	Japan	Vehicles	McDonald's	USA	Restaurants	57,000
7	Intel	USA	Computer hardware	Samsung Group	South Korea	Electronic conglomerate	55,000
8	McDonald's	USA	Restaurants	Facebook	USA	Social media	49,000
9	Disney	USA	Entertainment	Mercedes-Benz	Germany	Vehicles	47,000
10	Google	USA	Internet services	IBM	USA	Computer services	46,000

Power is another example of a fast-changing geographical theme – it ebbs and flows and continually changes.

Table 2.9 is an example of 'subjective' data. The writer researched widely and came to this personal summary. Each country has been given a value out of five, indicating the forms and significance of six forces of power: power reach, military, economic, cultural, colonial-based and media and information. You may disagree with these values and that's to be expected and even encouraged. In Geography, information comes from many sources. Sometimes it is simply based on the views of a person or a group, so it may be biased or even untrue. You need to be able to distinguish between subjective views and objective views.

Your research:

- How can power 'ebb and flow'?
- Explain the changing nature of economic power for individual countries such as China, the UK or India.
- From Table 2.8, select one or more of the global brands and investigate its global growth and success.

▼ Table 2.9 Sources and forms of power for selected countries/regions

	Power reach	Military power	Economic power	Cultural power	Colonial-based power	Media and information power
USA	Global	5	5	5	1	5
China	Global	4/5	4/5	3	2	4
European Union (group of countries)	Global	3	5	4	3	4
Japan	Regional	1	3	3	1	1
Brazil	Regional	2/3	3/4	3	2	2
Russia	Increasingly global	4	3	2	2	2/3
UK	Declining global	2	3/4	4	4	4
India	Increasingly global	3/4	4	3	2	3

Case study: differences in levels of development between two developing countries, Ethiopia and Nepal

> ### Research opportunity
>
> Compile a background database for both countries including a summary under these selected headings:
>
> - Location
> - Population size and density
> - HDI and PQLI indices
> - Economy and development
> - Vulnerability to natural hazards
> - Debt, trade and aid
> - Political tensions and internal disputes

Nepal is a landlocked, mountainous country in South Asia, bordered by countries such as China and India. It is ranked 149th in the world in terms of HDI and, when measured by most social, political and economic indicators, it is clearly economically deprived. Nepal contains several different ethnic and religious groups (mostly Buddhists and Hindus) and political viewpoints (Maoists, communists, monarchists) but over the last 40 years the country has been caught up in conflict. Over 10,000 people have been reported killed in a series of civil wars.

Nepal has eight mountains of over 8000 metres and remains a magnet for serious mountaineers. The UK Foreign Office suggests that British people may visit Nepal with caution, with over 50,000 visiting in 2017. Its advice is: 'Never trek alone. Use a reputable agency, remain on established routes, and walk with at least one other person ... Altitude sickness is a risk ... You should take note of weather forecasts and conditions.'

Young Nepalese migrate overseas for employment (especially to India and the Middle East) although within the country there are several refugee camps caught up in a dispute with Bhutan.

Money enters Nepal from the UK (a legacy of many Nepalese Gurkhas who fought in the British army) and remittance from migrant workers sending money back home.

Ethiopia is located in north-east Africa (bordered by Kenya, Sudan, Somalia, Djibouti and Eritrea) and, with a population of 90 million, it is the most populous landlocked country in the world. The landscape is varied, with vast mountain ranges, tropical forest, semi-deserts, fertile farmland and many river systems. The country has over 80 ethnic groups and continues to be caught up in internal disputes. Although Ethiopia has vast water resources (14 major rivers, including the Nile), it remains economically deprived. However, agricultural output

▲ **Figure 2.16** Nepalese migrant construction workers in Qatar for the 2022 World Cup

is increasing (80 per cent of the exports) and since 2011 economic growth has been considerable, led by exporting quality leather goods and cut flowers.

Ethiopia's potential is considered significant due to its water and HEP resources and the possibility of further oil and gold exploitation.

 ## Task

1 The percentage of adult literacy is a social indicator of development. Explain how this indicator might illustrate a country's level of development.

▼ **Table 2.10** Adult literacy rates, 2018

Country	% adult literacy
Afghanistan	32
Bolivia	93
Kenya	79
Cuba	100
Ethiopia	39

2 Look at Table 2.10 and suggest reasons why such differences exist between developing countries.
3 Using Figure 2.17 as a guide, comment on the relationship that appears to exist between wealth (GDP/capita) and the percentage of adult literacy rates.

4 Nepal and Ethiopia are two of the least developed countries in the world. Look at the information in Table 2.11 and the information in the notes for each country in Figures 2.18 and 2.19 on page 60. Describe and explain the differences in levels of development between Ethiopia and Nepal.

▼ **Table 2.11** Development indicators for Ethiopia and Nepal, 2018

Development indicator	Ethiopia	Nepal
GDP per capita (US$)	550	728
Doctors per 100,000 people	10	65
Adult literacy (%)	39	60
Infant mortality rate per 1000 live births	41	28
Life expectancy (years)	66	71
HDI value	0.463	0.574

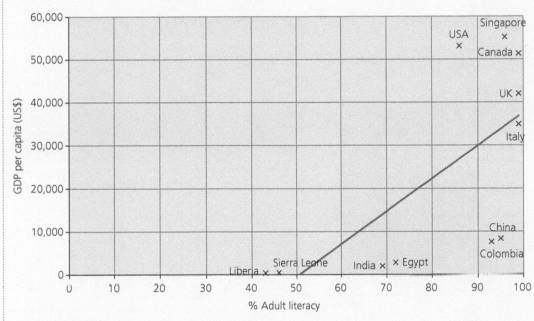

▲ **Figure 2.17** Scatter graph showing the correlation between adult literacy rates and GDP per capita (US$)

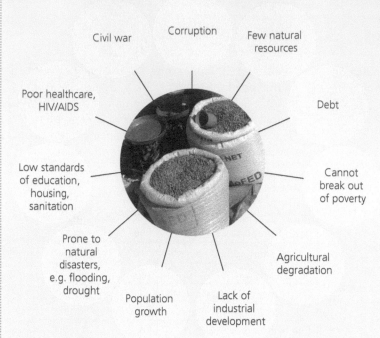

Civil war

Corruption

Few natural resources

Poor healthcare, HIV/AIDS

Debt

Low standards of education, housing, sanitation

Cannot break out of poverty

Prone to natural disasters, e.g. flooding, drought

Agricultural degradation

Population growth

Lack of industrial development

▲ **Figure 2.18** Background notes, Ethiopia

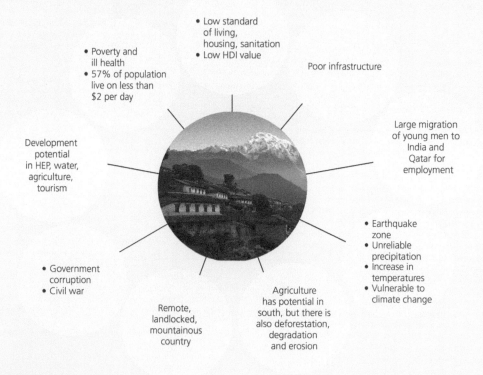

• Poverty and ill health
• 57% of population live on less than $2 per day

• Low standard of living, housing, sanitation
• Low HDI value

Poor infrastructure

Development potential in HEP, water, agriculture, tourism

Large migration of young men to India and Qatar for employment

• Government corruption
• Civil war

• Earthquake zone
• Unreliable precipitation
• Increase in temperatures
• Vulnerable to climate change

Remote, landlocked, mountainous country

Agriculture has potential in south, but there is also deforestation, degradation and erosion

▲ **Figure 2.19** Background notes, Nepal

Conclusion

A positive final statement from the 2013 Human Development Report:

> 'The 21st century is witnessing a profound shift in global dynamics, driven by the fast-rising powers of the developing world. China has overtaken Japan as the world's second biggest economy, lifting hundreds of millions of people out of poverty in the process. India is reshaping its future with new entrepreneurial creativity and social policy innovation. Brazil is raising its living standards by expanding international relationships and anti-poverty programmes that are emulated worldwide. The "Rise of the South" is a much larger phenomenon. Indonesia, Mexico, South Africa, Thailand, Turkey and other developing countries are becoming leading actors on the world stage. The 2013 Human Development Report identifies more than 40 developing countries that have done better than expected in human development in recent years, with their progress accelerating markedly over the last ten years.'

Source United Nations

However, despite the positive message in the UN's Human Development Report, it is clear that there is still work to do.

- In regions such as sub-Saharan Africa, little progress has been made and the gap has widened.
- The global recession from 2008 may have tipped up to 200 million people back into poverty and hunger, reversing progress made so far.
- Corruption, conflict and bad governance still affect many of the LDCs.

There is some good news, however.

- In some countries such as India and China, the gap between the rich and poor has narrowed.
- Debts have been reduced for some of the less developed countries, giving them the chance of a new start.
- Aid in 2013 was at a record level.
- Initiatives such as Fair Trade have had a positive impact.

- Progress has been made with diseases such as HIV/AIDS and malaria.

You have covered a lot in this section on development and should now be able to answer questions on:

- social, political, economic single and composite indicators
- differences in levels of development between developing countries.

2.5 Introduction to health

In this part of the chapter we will consider how levels of health and the incidence of disease are major indicators of development. These illustrate the ability of a society to cope with ill health and the environmental, social, human and lifestyle conditions that can be causal or worsen the problems of poor health.

For the SQA course and exam the key areas focused on are:

- a water-related disease: causes, impact, management
- primary healthcare strategies.

To be successful you will need to be able to demonstrate the interaction of physical and human factors and evaluate strategies adopted in the management of the global health issue studied.

As with most of the global issues studied, there is an element of choice. For health it is the decision of which water-related disease to study. According to the World Health Organization (WHO), water-related diseases include:

- those due to micro-organisms and chemicals in the water people drink
- diseases like schistosomiasis which have part of their life cycle in water
- diseases like malaria with water-related vectors
- others such as Legionellosis carried by aerosols containing certain micro-organisms.

For the purposes of this book we will focus on malaria, which is recognised as the world's most important parasitic infectious disease. It is a water-related disease as it is transmitted by mosquitoes which breed in water.

2.6 Factors influencing health

An individual's health is related to hereditary and biological factors, as well as environmental, physical, social and human conditions. The factors that influence health can be divided into physical and human factors (Table 2.12).

▼ **Table 2.12** Factors influencing health

Physical factors	Human factors
● Climate conditions (leading to drought, flooding, crop failure, famine – all positive conditions for diseases and their vectors to thrive) ● Availability of clean fresh water ● Presence of **endemic disease** ● Remoteness/ inaccessibility/ poor communications (reduces access to even basic medical care) ● Famine	● Poverty (unable to afford healthcare or high-quality, nutritious food) ● Poor living conditions (e.g. shanty towns) ● Overcrowded conditions (diseases spread quickly) ● Lack of sanitation or poor sanitation ● Lack of access to clean water ● Insufficient healthcare support, hospitals, drugs ● Poor education, low public awareness of causes, prevention methods or effective treatments ● Poor hygiene

It is worth noting that more than a billion people live in conditions that deny them clean water, adequate food (amount and/or quality), basic healthcare or education.

2.7 Malaria: an introduction

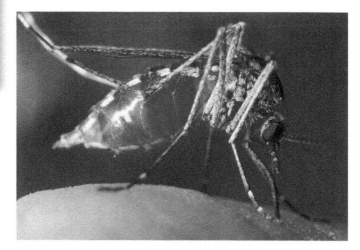

▲ **Figure 2.20** Female *Anopheles* mosquito

Malaria is a life-threatening infectious disease that is transmitted to humans and animals by the female *Anopheles* mosquito (Figure 2.20 on page 62). The mosquitoes are infected with microscopic malaria parasites which they have received when biting an already infected human or animal and ingesting infected blood. The mosquito then infects other humans or animals by biting them and injecting the parasites into their bloodstream.

Malaria is mostly found in the tropical and semi-tropical areas of the world (Figure 2.21). The climate conditions within these areas are suitable for the parasites to survive and for mosquitoes to thrive:

- temperatures above 16°C and below 30°C
- precipitation levels that allow for stagnant, surface water accumulation which acts as mosquito breeding grounds
- 26–60 per cent humidity.

Altitude can be a component of this, with areas of malarial risk tending to be below 3000 m.

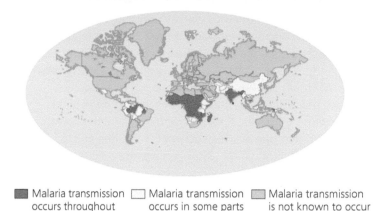

■ Malaria transmission occurs throughout ☐ Malaria transmission occurs in some parts ☐ Malaria transmission is not known to occur

▲ **Figure 2.21** Malaria: global locations

The availability of areas of stagnant water suitable for the *Anopheles* mosquito eggs to be laid is a crucial factor in allowing the mosquitoes to survive and become effective and numerous vectors for the disease. The mosquito needs water in which to lay its eggs. This is followed by a period of 5 to 14 days of development in their life cycle from egg to larva and then pupa. Following this, the mosquito is finally released and enters its flying stage as an adult (Figure 2.22).

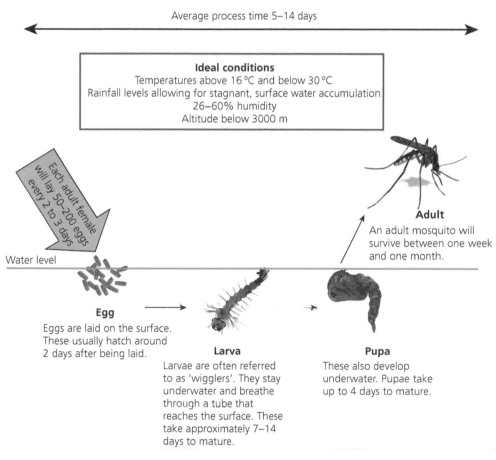

Average process time 5–14 days

Ideal conditions
Temperatures above 16 °C and below 30 °C
Rainfall levels allowing for stagnant, surface water accumulation
26–60% humidity
Altitude below 3000 m

Each adult female will lay 50–200 eggs every 2 to 3 days

Water level

Adult
An adult mosquito will survive between one week and one month.

Egg
Eggs are laid on the surface. These usually hatch around 2 days after being laid.

Larva
Larvae are often referred to as 'wigglers'. They stay underwater and breathe through a tube that reaches the surface. These take approximately 7–14 days to mature.

Pupa
These also develop underwater. Pupae take up to 4 days to mature.

▲ **Figure 2.22** *Anopheles* mosquito from egg to adult

Anopheles mosquitoes will lay their eggs in a variety of sizes and types of water areas. They are found in fresh and saltwater marshes, rice fields (paddy fields), undisturbed edges of streams and rivers, mangrove swamps, puddles and other rain pools. In addition to these, tin cans, holes in trees, ornamental ponds, swimming pools, flooded tyre tracks, ditches, sewage channels, barrels, buckets, water troughs and virtually any still or stagnant water can provide a suitable location.

Malaria affects over 90 countries and some 27 per cent of the Earth's land surface. According to recent WHO findings, this puts almost half the world's population at risk. Of all malaria deaths, 90 per cent occur across the continent of Africa, where over 400 million people live in areas where malaria is endemic. Sub-Saharan Africa has climatic conditions that favour mosquito proliferation and this, combined with poor prevention and treatment, has made malaria one of the biggest causes of infant mortality.

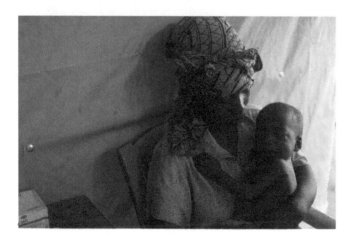

▲ **Figure 2.23** Young child suffering from malaria

Figures for 2016 collected by the WHO show that there were 216 million cases of malaria worldwide, resulting in up to 445,000 deaths. Children under the age of five are more susceptible to infection and illness and, as such, they make up 70 per cent of all malaria deaths. It is estimated that one child aged under five years died from malaria every two minutes in 2016. However, there have been successes in malaria treatment and prevention programmes, reducing deaths due to the disease. The number of deaths of under-fives due to malaria has been reduced from 444,000 to around 285,000 in the years from 2010 to 2016.

 Task

1 What are the climate conditions that best suit the survival of the malarial parasite and the *Anopheles* mosquito?
2 Describe, and give examples of, the condition and types of water areas that are suited for mosquitoes to lay their eggs.
3 In some areas malaria is described as being endemic. What does this mean?
4 At present malaria is found in approximately 27 per cent of Earth's land area. How many people are at risk from malaria? Answer in number form.

Causes of malaria

Malaria in humans is caused by five species of the *Plasmodium* parasite (single-cell micro-organisms) (Figure 2.24):

- *P. falciparum*
- *P. vivax*
- *P. ovale*
- *P. malariae*
- *P. knowlesi.*

Each of these types of *Plasmodium* parasite has its own specific effects, creating different symptoms or severity. They can also be found in different locations. *P. knowlesi* has only recently been acknowledged as causing malaria in humans. This was previously thought to affect only macaque monkeys but is now believed to have infected humans in south-east Asia and, in particular, Borneo.

Human malaria is only transmitted by the female *Anopheles* mosquito (Figure 2.20). There are 430 *Anopheles* species and, of these, between 30 and 40 are vectors for malaria. The male of this species does not transmit the disease. The female mosquito needs human blood to nourish the eggs that it lays (50–200 every 2 to 3 days) and this makes the mosquito very active in its search for blood. Mosquitoes are believed to identify the person they will bite by the odours given off, the individual's skin chemistry and through some visual clues. One person in ten is especially attractive to mosquitoes.

Studies have shown that mosquitoes will enter houses or sheltered/shady areas during the early morning and during the evening. They may hide in cupboards, between floorboards, in clothing, curtains or in any other dark, cool corners and spaces. The highest level of biting activity for malaria-transmitting

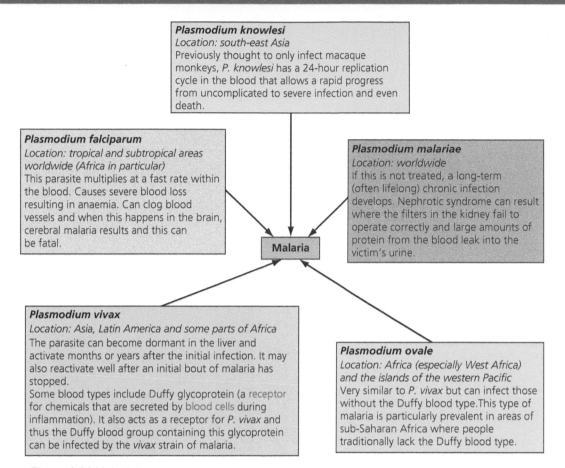

Plasmodium knowlesi
Location: south-east Asia
Previously thought to only infect macaque monkeys, *P. knowlesi* has a 24-hour replication cycle in the blood that allows a rapid progress from uncomplicated to severe infection and even death.

Plasmodium falciparum
Location: tropical and subtropical areas worldwide (Africa in particular)
This parasite multiplies at a fast rate within the blood. Causes severe blood loss resulting in anaemia. Can clog blood vessels and when this happens in the brain, cerebral malaria results and this can be fatal.

Plasmodium malariae
Location: worldwide
If this is not treated, a long-term (often lifelong) chronic infection develops. Nephrotic syndrome can result where the filters in the kidney fail to operate correctly and large amounts of protein from the blood leak into the victim's urine.

Malaria

Plasmodium vivax
Location: Asia, Latin America and some parts of Africa
The parasite can become dormant in the liver and activate months or years after the initial infection. It may also reactivate well after an initial bout of malaria has stopped.
Some blood types include Duffy glycoprotein (a receptor for chemicals that are secreted by blood cells during inflammation). It also acts as a receptor for *P. vivax* and thus the Duffy blood group containing this glycoprotein can be infected by the *vivax* strain of malaria.

Plasmodium ovale
Location: Africa (especially West Africa) and the islands of the western Pacific
Very similar to *P. vivax* but can infect those without the Duffy blood type. This type of malaria is particularly prevalent in areas of sub-Saharan Africa where people traditionally lack the Duffy blood type.

▲ **Figure 2.24** Malaria types

mosquitoes is experienced from around midnight and into the early hours of the morning.

The female mosquito has adapted so that its bite penetrates skin and retrieves blood (Figure 2.25). It has a long, thin **proboscis** (an extended tubular mouthpiece used for feeding and sucking). Inside the proboscis there are six needles which help in the biting process. Two of these needles have sharpened cutting edges that slide across and cut the skin. The next two hold the cut and other tissues apart. Another needle has receptors that identify the chemicals given off by blood vessels and this directs the proboscis to the blood. The same needle is then used like a straw to drink the blood, while the final one injects saliva into the wound. This saliva suppresses clotting of the blood, allowing it to flow easily, and also suppresses any pain.

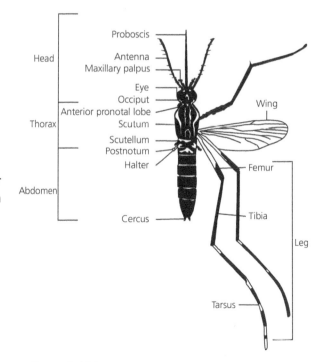

▲ **Figure 2.25** Diagram of a female *Anopheles* mosquito

Life cycle of the malarial parasite

If the host is infected with malaria, the blood ingested by the mosquito will contain the parasites. The malaria parasites then complete their growth cycle within the mosquito. The completion of this cycle will take from 10 to 18 days, depending on the conditions at the time: higher temperatures and humidity accelerate the process. Temperature is also critical, depending on the type of *Plasmodium*. *P. falciparum* cannot complete its growth cycle if temperatures are below 20°C and in these conditions malaria would not be transmitted.

The parasite must complete its development within the mosquito before it becomes infectious to humans. With the process completed, the parasite is located within the mosquito's salivary gland. Now when the mosquito bites and introduces its saliva into another person's blood, the saliva contains the malaria parasites and will cause infection.

Once within the bloodstream the parasites migrate to the host's liver where they mature and reproduce before returning to the bloodstream. The parasites then penetrate the red blood cells and continue to multiply. This eventually causes the red cells to burst

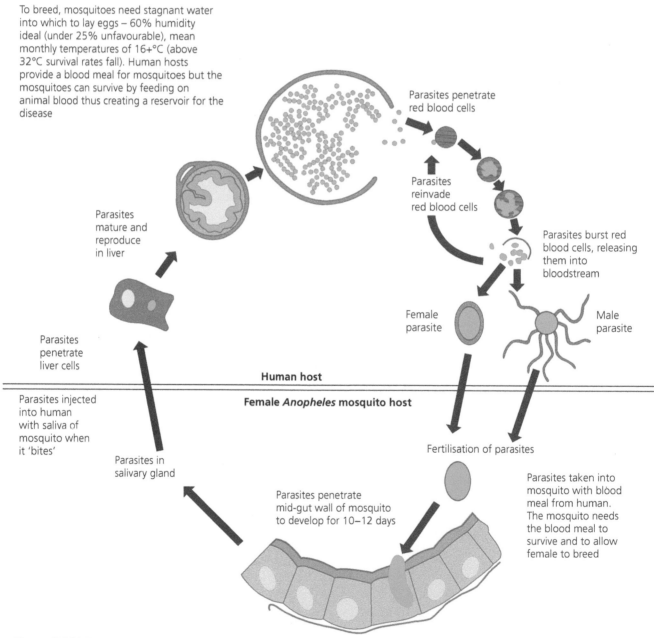

To breed, mosquitoes need stagnant water into which to lay eggs – 60% humidity ideal (under 25% unfavourable), mean monthly temperatures of 16+°C (above 32°C survival rates fall). Human hosts provide a blood meal for mosquitoes but the mosquitoes can survive by feeding on animal blood thus creating a reservoir for the disease

Parasites penetrate red blood cells

Parasites reinvade red blood cells

Parasites mature and reproduce in liver

Parasites burst red blood cells, releasing them into bloodstream

Parasites penetrate liver cells

Female parasite

Male parasite

Human host

Parasites injected into human with saliva of mosquito when it 'bites'

Female *Anopheles* mosquito host

Parasites in salivary gland

Fertilisation of parasites

Parasites penetrate mid-gut wall of mosquito to develop for 10–12 days

Parasites taken into mosquito with blood meal from human. The mosquito needs the blood meal to survive and to allow female to breed

▲ **Figure 2.26** Life cycle of the malarial parasite

and release free parasites into the blood plasma. Some of these will also infect other red blood cells and multiply further until the cells burst in the same manner as before.

When the parasites are released into the blood plasma, this causes the person to experience the fever associated with the disease. The free parasites in the plasma now become the source of infection for other mosquitoes that feed on the host.

If treated quickly enough, and in the correct manner, malaria is usually curable.

Task

For this task you may like to create an information diagram or poster, a mind map, a spider graph or a PowerPoint/media presentation. What is important is to create a good learning tool for yourself or your class.

Describe the process of malarial infection using the headings below:

- The vector
- How the vector bites
- How the vector becomes 'infected'
- Conditions for laying eggs
- Development cycle of the mosquito and the parasite
- How the vector infects a human
- How the parasite develops inside the host

Symptoms of malaria

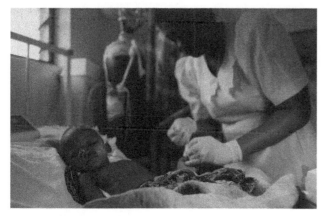

▲ **Figure 2.27** Nurse treating a malaria patient

In general, the symptoms of malaria include a headache, swinging violently between fever and shaking chills and an enlarged spleen. This is

accompanied by muscle aches, sweats, extreme tiredness, nausea and vomiting. Malaria sufferers may also develop anaemia and jaundice (where skin and eyes turn yellow) because of the loss of blood cells caused by the red cells bursting after the parasites multiply within them.

Following the bite there is normally an incubation period before the symptoms appear. This is usually between 7 and 30 days, although some examples have been up to a year. *P. falciparum* commonly takes the shortest time and *P. malariae* the longest. *P. vivax* and *P. ovale* infections add an extra complication because the parasites can lie dormant in the victim's liver for up to four years and activate/reactivate at any time within that period.

People can contract different levels of severity of malaria and these can be categorised as uncomplicated or complicated (severe). As we have already noted, each of the different malarial *Plasmodia* can result in different symptoms, severity and complications. For some people, repeated infection can allow them to develop a protective immunity while for others there may be acute repercussions.

When malaria is extremely severe (especially with *P. falciparum*), it can result in organ failure and even extreme post-malarial effects (Figure 2.28 on page 68).

Additional factors in the spread of malaria

Sometimes when learning about a disease for an exam it is easy to concentrate on the numbers, graphs and economic statistics and forget the personal impact. For individuals, malaria is a terrifying, debilitating, painful, quality-of-life ruining, death-inducing curse. Families suffer and can fall deeper into poverty trying to survive when even just one of them is infected. Where people rely on their own efforts to grow crops, raise animals, gather food or simply collect water, this disease can make everyday activities so much more difficult and at times impossible. Hunger, thirst, weakness and susceptibility to further diseases and infections become additional outcomes.

Malaria does not only affect those from developing areas. It is a result of climate and environmental conditions and has the ability to affect anyone who is unprepared. What is certain is that poverty is a major factor which increases the possibility of becoming infected and being unable to find or afford treatment.

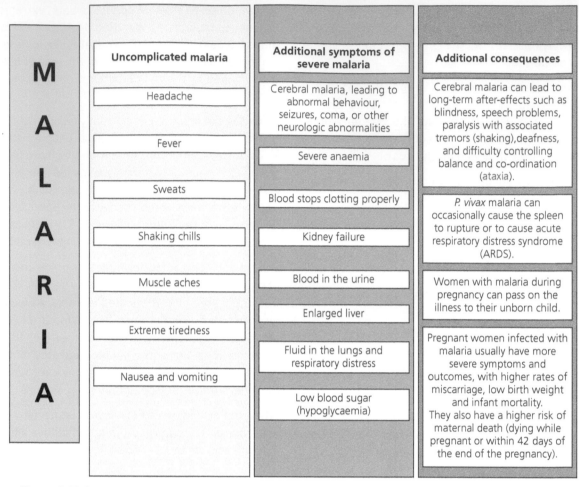

▲ **Figure 2.28** Symptoms and consequences of malaria

This has knock-on effects by providing another host and source for the malaria parasite, creating increased risk for others.

Individuals and communities can be put at risk due to traditional beliefs, social understandings, economics and level of education. Some of this is due to the need to survive and manage with limited sources of food and water. Risk factors for malaria infection are described below.

- Traditional housing construction with wood/branches/mud bricks or with open vents and windows allows mosquitoes to enter as such houses cannot be fully sealed.
- Living close to pools or other areas of still water in order to access them for a water supply or as a source of food (e.g. fish) places communities near to mosquito breeding grounds and increases risk.
- Living in areas where sewage is not adequately removed (such as in shanty towns) provides stagnant ponds which attract mosquitoes.

- Agricultural work puts individuals at risk, especially during harvest time, if the activity takes place later into the evening. Due to the warm climate conditions it is more likely that a worker will not be fully covered and will leave areas of skin exposed to the possibility of insect bites.
- Paddy fields (rice fields) create perfect egg-laying areas due to their flooded nature.
- Digging irrigation ditches, or ploughing furrows, can create areas for standing water in which the mosquito can lay its eggs and the larvae develop.
- Although rearing animals may provide an alternative blood source for the *Anopheles* mosquito which could decrease human exposure, it is also possible that by working with these animals a person will have increased exposure to the mosquitoes.

- Being from an economically deprived area can mean that even the simplest of anti-malaria prevention or treatment is unaffordable.
- With limited education or information, the causes and symptoms of malaria may not be understood and this can lead to poor responses and a lack of, or incorrect, treatment.
- Cultural beliefs may result in the use of traditional but ineffective methods of treatment.

So that a family can survive when parents or other siblings are infected, children may be expected to carry out additional work and be removed from any education which they have. They may then be unintentionally forced to go to the areas where other members of the family were exposed to malaria while carrying out their daily tasks.

Due to attempts to help other family members with malaria, families will have reduced time and financial ability to invest in aspects of their children's health, welfare or education.

Malaria may not be caused by poverty, but poverty makes infection more likely and assists in the embedding and spreading of the disease.

Impact of malaria

It is estimated that around 50 per cent of the world's population lives within areas that suffer from malaria. Malaria places a massive burden on the economies and development of countries where the disease is rife. These countries have to spend a large amount of their income coping with the disease instead of investing in improvements to their general economy. Malaria affects the working ability and productivity of its people and a lack of investment severely restricts a country's level of economic development over time. Where there is a high disease rate, the economic impact results in a much lower GDP than a country without malaria. It has been calculated that malarial endemic countries have a 1.3 per cent smaller growth rate and that over a 15-year span GDP is reduced by 20 per cent. If malaria is reduced by 10 per cent, it has been shown that this would result in a 0.3 per cent increase in a country's growth rate. In total, it is estimated that the continent of Africa loses around £7.7 billion ($12 billion) every year due to the impact of malaria.

Malaria can affect a country's economy in a number of ways:

- Work absenteeism increases, thereby reducing productivity and the reliability of supply to customers.
- Crop production is affected either due to:
  - absenteeism or inability to put in effective levels of effort, or
  - farmers planting subsistence crops rather than more labour-intensive cash crops as they have a reduced ability to look after them.
- Foreign investment is inhibited because:
  - company employees are unwilling to travel to malaria endemic areas (either to source, investigate or take part in management)
  - companies are unwilling to invest if products/production cannot be guaranteed.
- Tourism is severely restricted as travellers are reluctant to visit. As a result, the benefits linked to tourism (such as foreign currency, building of facilities, infrastructure projects, agricultural sales, employment and local and national economic boosts) do not happen.
- Increases in school absenteeism, through illness or pupils having to provide for themselves or their families, reduces the level of education of the working age group and so limits possible industrial or technological advances within the nation.
- Deforestation, road construction and agricultural development which are intended to boost a country's economy can actually increase the intensity of malaria transmission, since all of these developments may create places for water to collect and increase mosquito breeding grounds. Increased numbers of people moving into or through an area also means additional potential hosts for the parasite. This may not only increase the number of people with malaria but also encourage its spread.

Lack of economic development has an additional effect in that the country has a reduced ability to tackle malaria in the first place. The WHO has identified stress factors on the ability of less developed countries, and their often under-provided or limited health systems, due to endemic malaria. The disease accounts for:

- up to 40 per cent of all spending on public health
- 30 to 50 per cent of all inpatient admissions to hospitals
- up to 60 per cent of all outpatient visits to health clinics.

As noted earlier, malaria is not caused by poverty although having the financial ability to tackle it does improve conditions notably. In 1993, the Sultanate of Oman had a GDP of £6666 ($10,000) per person and was often referred to as an example of an ostensibly rich nation which still had endemic malaria. However, by using the finances available to it (due to its oil-related wealth) and implementing healthcare strategies that included education, additional medical staff and training, recording and reporting systems, dealing with mosquito breeding grounds and setting up treatment centres and supplies, the Sultanate of Oman managed to reduce malarial infections from more than 30,000 cases in the early 1990s to 1087 in 2017.

 Task

1 What are the human factors that put people at risk of malaria?
2 What changes would happen to the lives of individual people if malaria were to be eradicated?
3 What would be the benefits of controlling or eradicating malaria to a developing country as a whole?

Management of malaria

In ancient times, people struggled to deal with malaria and more recently the battle has been to manage or eradicate the disease. WHO efforts to eradicate malaria worldwide in the latter half of the twentieth century failed due to numerous flaws in the strategy, including:

- a very strict approach that was resisted by some countries
- failure to consider the ability of nations to afford the strategy
- lack of infrastructure and finance, meaning some countries were unable to access all their regions
- not considering local conditions or needs
- not adapting the approach to the different types of malaria
- omitting large areas of endemic malaria (e.g. sub-Saharan Africa)
- being unprepared for developing resistance to the insecticide DDT and drugs
- basing the approach on DDT, which some countries could not afford (along with the aircraft and spraying equipment needed to cover large

areas), and which was harmful to the environment and humans, interfering with reproductive health and acting as a carcinogen
- occurrence of wars and resultant large-scale population movements, making access to areas and control of the spread of the parasite almost impossible
- reduction in funding.

The attempt to eradicate malaria was suspended in the late 1960s; during the 1970s and 1980s malaria cases and resulting mortality increased at an extreme rate. New approaches, developed by learning from previous mistakes, were needed.

The Roll Back Malaria initiative (WHO, 1998) defined targets for malarial control and highlighted the need for much improved *local* health systems to make sure any action was sustainable and supported. Increased funding was also made available through the Global Fund to Fight AIDS, Tuberculosis and Malaria, various international sources and the World Bank.

In 2000, the United Nations Millennium Development Goals (MDGs) highlighted malaria as one of the biggest barriers to global development. One of the targets set was to halt and begin to reverse the occurrences of malaria by 2015.

The WHO approach from 2000 tackled some of the problems of early anti-malaria plans by incorporating:

- sustainability in areas with very limited resources
- the use and strengthening of local capacities so that sustainability is built in when UN, WHO or other specialists leave
- consideration of local needs and situations
- a better understanding of malaria transmission
- better co-ordination between global bodies, governments, charities, researchers, and both public and private groups
- improved and guaranteed levels of funding
- increased research and development
- quicker transmission of information and the sharing of research findings and reports, allowing faster response times and more focused and practical research
- early detection, containment and prevention of epidemics.

In 2008 the WHO announced that a drive for the eradication of malaria was 'back on the global health agenda'.

It has become clear that large amounts of money are needed to provide and transport materials and drugs, and to fund anti-malaria strategies, research and development. Global funding increased by 1000 per cent between 2000 and 2018 due to a co-ordinated approach, with international agreements negotiated by the UN and WHO resulting in long-term pledges from countries under threat, wealthier unaffected countries round the world, banks, and private and commercial interests.

The Millennium Development Goal to 'halt and begin to reverse the occurrences of malaria by 2015' was generally successful. There had been a marked decline since 2000 but figures showed a concerning up-rise in reported cases in 2014–15, which continued into 2016. The period 2015–16 saw a flattening out in the reduction in deaths and in 2017 there was a steep increase. It was generally accepted that this was due to a growing resistance to the insecticides and drugs used to fight malaria (Figure 2.29).

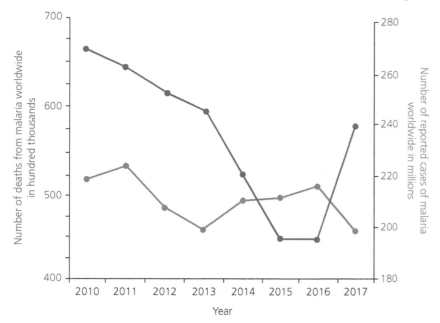

The figures shown are taken from the annual reports of the WHO. There are other sources for figures available, but most show similar trends for the period reported.

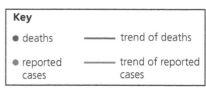

Key

● deaths   ——— trend of deaths

● reported   ——— trend of reported
cases   cases

▲ **Figure 2.29** Deaths and reported cases of malaria, 2010–17

In 2015, the WHO *Global Technical Strategy for Malaria 2016–2030* set new targets:

- by 2030, to reduce malaria incidence by at least 90 per cent
- by 2030, to reduce malaria mortality rates by at least 90 per cent
- by 2030, to eliminate malaria in at least 35 countries
- to prevent a resurgence of malaria in all countries that are malaria-free.

In the period between 2000 and 2018, there were marked successes:

- Deaths from malaria were reduced by 60 per cent globally, with around 7 million lives saved.
- There was a 25 per cent decline in new cases of malaria.
- Malaria mortality rates decreased by around 29 per cent since 2010, due to increased prevention and control measures.

- Mortality rates in the under-five age group were reduced by around 35 per cent since 2010.
- 57 countries with malaria transmission in 2000 reached the target of a 75 per cent reduction in malaria cases by 2015 (WHO target set in 2005).
- Between 2007 and 2018, eight countries were certified by the WHO as having eliminated malaria, and five more were in the process of applying for certification. Those who were certified malaria-free were: United Arab Emirates (2007), Morocco (2010), Turkmenistan (2010), Armenia (2011), Maldives (2015), Sri Lanka (2016), Kyrgyzstan (2016) and Paraguay (2018).

Bill Gates, of the Bill & Melinda Gates Foundation, saw the successes as worthy but said that much was still to be done, especially due to the apparent increase in resistance to drugs and insecticides. Gates emphasised the need for funds to allow research and development to keep ahead of resistance and new forms of malaria. He warned that, even in areas of present success, 'if we don't keep innovating, it will return'.

Gates highlighted that a variety of tools were needed to tackle malaria. He also stated that surveillance, identifying, reporting, data collection and speedy response were extremely important. The development of low-cost, easy-to-use diagnostic tools allows *local* people to quickly detect malarial infection. These diagnostic tools are now as small as a finger and need only a tiny drop of blood to identify infection. In the past, blood tests had to be sent to a specially equipped laboratory and results confirmed by a trained scientist, taking weeks.

Advancements in communications have resulted in simple hand-held GPS systems (using **satellites**) that can show locations of outbreaks in real time, with mobile phones facilitating instant reporting and requests for assistance. Computers can quickly analyse data and help create visualisations to allow better understanding of the disease and faster decision making.

Drones are used to help identify locations of mosquito breeding. They can provide quicker and more precise readings than satellites. Satellite imagery can take weeks to become fully available, but photographs and sampling by drones can be available on the same day. This allows a much faster response, for example the prompt spraying of breeding areas with insecticides. Drones have also been used to move supplies, equipment and medicines to areas with people in need.

One of the greatest steps forward, and one with the potential to vastly reduce the incidence of malaria, has been the move towards a successful **vaccine**. Many scientists believe that only an effective anti-malarial vaccine can bring about a large-scale change in the fight against the disease in endemic areas. It is hoped that a vaccine will supplement all the other anti-malarial efforts already in use. Several vaccines are under trial but, due to the complexity and the different varieties of malarial parasite, vaccine development is extremely difficult.

One vaccine – RTS,S (Mosquirix) – which provides partial protection against *Plasmodium falciparum*, entered the WHO-approved, large-scale implementation study stage in 2018. Previous smaller-scale trials showed that RTS,S had the potential to reduce malaria by 26 to 50 per cent in infants and young children. The large-scale implementation studies in Ghana, Kenya and Malawi have 750,000 children taking part. It is hoped that the vaccine will be fully certified for use by 2023.

Recent scientific advancements, allied to the increase in funding for research, has allowed for progress in **gene drive** approaches to malaria eradication. Gene drive is **genetic engineering** technology that can spread a type or set of genes throughout a species. The technique can add, remove, modify or disrupt genes so that the species exhibits qualities which are wanted. To reduce malaria, gene drive research is targeting the following:

1   Interfering with mosquito survival or reproduction:
    ● so that males become infertile and breeding is impossible (causing the population to reduce and eventually die out)
    ● to introduce a gender bias towards males so the number of females available to act as vectors is reduced
    ● to genetically arrange for female mosquitoes to die if they become infected with the mosquito parasite.
2   Altering the target for transmission:
    ● to create a *dislike* for humans as a target for feeding (by triggering a reaction against human hormones) and force feeding onto non-humans.
3   Neutralising the mosquito as a vector:
    ● to modify the female mosquito so that it is unable to carry the parasite
    ● to interfere with parasite development within the female mosquito so that the parasite either dies or becomes unable to infect.

The introduction of vaccines and gene drive techniques are seen as possible game changers in the fight against malaria (Figure 2.30 on page 73).

Within its modern scientific approach (Figure 2.31 on page 73), the WHO suggests that two relatively cheap and simple strategies are effective as a *basis* for malaria prevention in all areas. These are insecticide-treated mosquito nets and indoor insecticide spraying. Both greatly assist in reducing the contact between mosquitoes and humans.

What is certain is that no one solution has been found and that future methodologies will need to employ a *mixture* of strategies to deal with the conditions/type of parasite in the areas where malaria is being tackled.

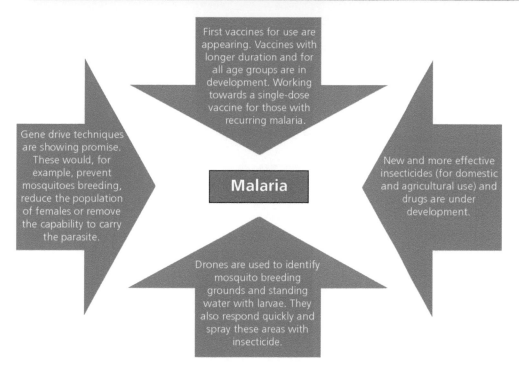

▲ **Figure 2.30** New tools for a modern approach to malaria

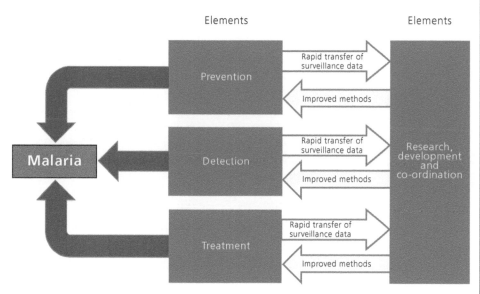

This diagram shows the basis of most modern approaches to malaria control.

There are four main elements, all of which can involve use of a number of tools, and which relate to the local need or situation.

Rapid transfer of detailed surveillance data allows a faster and more effective response from research and development teams (e.g. earlier notification of resistance to insecticides and drugs and the creation of new effective insecticides and drugs), cutting the lead time to new developments and potentially saving lives.

Rapid transfer of detailed surveillance data also allows swifter co-ordination and allocation of resources to deal with outbreaks and changing needs in different areas.

All the elements of this model need continuous, large-scale and long-term funding to operate effectively.

▲ **Figure 2.31** Modern co-ordinated approaches to malaria control

Some of the approaches to the control and eradication of malaria are shown and evaluated in Table 2.13 on page 74.

▼ **Table 2.13** Methods of approaching the control and eradication of malaria

Target/method	Reasons	Comment
Elimination of breeding areas		
Fill in and drain areas of standing/stagnant water	Without areas in which to lay their eggs, mosquito numbers will decline and the parasite will lose the ability to develop	Can be very costly and difficult to achieve. Mosquitoes can still lay their eggs in very small amounts of water, e.g. vehicle tracks, puddles, tin cans, bottles, animal tracks or hollows in trees. Except for large areas of water, this can be carried out by local communities or individuals. The drainage of swamps is more problematic and not always practicable in tropical areas as it takes a lot of effort and intense/regular rainfall can recreate the conditions relatively quickly.
Spray breeding and infested areas with insecticides	By killing the mosquitoes, there will be no vector for the parasite (i.e. nothing to transmit the disease)	To spray the large areas infected can be costly. Insecticides have to be paid for, as does the aircraft needed to cover widespread areas. Spraying by hand or from vehicles is time consuming, labour-intensive and there may be difficulties in accessing some areas due to lack of roadways or through difficult terrain. Some insecticides, such as DDT, can have harmful effects on humans, especially if they get into the food chain. DDT side-effects include: ● cancers ● male infertility ● miscarriages and babies of low birth weight ● developmental delay in children ● nervous system and liver damage. DDT is banned for agricultural use worldwide but is still used in some regions as the most efficient anti-malaria insecticide. It is now mostly used for indoor spraying. Certain oil-based insecticides have proved too expensive for some countries, but recent increases in funding have assisted with their purchase, as well as research, leading to the development of cheaper, more effective and often non-oil-based, targeted insecticides. Drones are being seen as a more effective, targeted and rapid method of spraying, with the advantage that local people can be trained to use them. *Anopheles* mosquitoes and other insects have shown that they can develop immunity to insecticides.
Spray oil on water	Causes the mosquito larvae to suffocate and drown	The oil can cause further damage to the ecosystem and even enter the food chain through fish. Diesel is used in preference in warmer climates as it eventually evaporates.
Flush out streams and rivers, e.g. by releasing water from dams	Washes away the eggs and larvae, drowns them and disturbs standing water	Works temporarily unless a continual flow of water is released. Water pools and standing water reappear after the water is stopped. Must be repeated every 7–10 days to disrupt the breeding cycle. Defeats the purpose of the dam to store water. Will not work in areas lacking the ability to control water.

▼ **Table 2.13** *(Continued)* Methods of approaching the control and eradication of malaria

Target/method	Reasons	Comment
Clear trees, shrubs and water plants (e.g. reeds) from streams	Removes shelter and shade for the mosquitoes and larvae. Mosquitoes rest in shaded areas to digest their blood meal. Removal of water plants makes it easier for fish and other predators to reach and eat the larvae	An effective method but it is also labour-intensive and time-consuming. It improves the possibility of a reduction of mosquitoes but does not eliminate the egg-laying areas. Except for very large areas, this can be carried out by local communities or individuals.
Plant eucalyptus	Eucalyptus absorbs large amounts of water and is very effective at draining marshes and swamps, thereby reducing mosquito habitats	With a little training and assistance in sourcing eucalyptus, local communities can carry out planting relatively easily. Over time eucalyptus trees can act as a vital source of timber as they grow and spread relatively quickly. This has caused a problem in South Africa where they are now being viewed as invasive, covering about 10% of the country, at the expense of native species. There is some concern that the trees soak up vital water from surrounding farmland.
Prevention of larvae development		
Introduce larvae-eating fish, e.g. Nile tilapia, muddy loach, mosquitofish	To remove the larvae before they can develop	Introduction of these fish is often difficult for some communities as they need to be purchased. National or international assistance is often needed. Major benefits are: • the fish breed and can be used elsewhere • they can be used in large areas such as ponds and lakes or even smaller ones such as irrigation ditches • they provide an additional food source (protein) for local communities. Damage to local ecosystems can be caused if fish that are introduced, e.g. the aggressive mosquitofish (*Gambusia affinis*) become predators to the indigenous aquatic life.
Bti (*Bacillus thuringiensis israelensis*) coconuts	Bti spores release toxins into the stomach lining of the larvae, which kills them	Bti is a naturally occurring soil bacterium that can kill mosquito larvae present in water. A hole is drilled into a coconut, Bti is inserted, sealed and left to ferment and the spores multiply. The coconut can be split open and the entire contents thrown into the infected water for the larvae to eat. Two or three coconuts can control a water area the size of a large pond for up to 45 days. In Peru it was shown that local communities could be taught the process and they were successful in carrying out the creation and use of Bti coconuts. This is a cheap method, particularly in areas where coconuts are plentiful. Bti is believed to be environmentally friendly, although some recent studies suggest that continuous application of Bti over a period of 2–3 years may result in a decrease in biodiversity. Bti can also be sprayed but this adds extra cost through the need for spraying equipment.

▼ **Table 2.13** *(Continued)* Methods of approaching the control and eradication of malaria

Target/method	Reasons	Comment
Drug treatments		
Use of drugs such as quinine, chloroquine, artemisinin and mefloquine	Used individually or in conjunction with other drugs, the main intent is to kill or stop the development of parasites following infection. Best results follow early detection	For many developing countries, drugs are very expensive and supplies are reliant on external aid. Such countries are unable to sustain an anti-malarial drug programme and so it becomes ineffective. The WHO warns that the efficacy of drugs being used needs to be strictly monitored as many forms of malaria have become resistant to them. Recent research funding has been put in place to try to ensure that resistance is identified early and new drugs developed. Some anti-malarial drugs have unpleasant side-effects. As a result, people being treated may not complete the entire course, helping to create resistance to the drug.
Prevention		
Spray the inside of houses with insecticides  ▲ **Figure 2.32** Indoor residual spraying (IRS)	**Indoor residual spraying (IRS)**	Insecticides such as Malathion are oil-based and expensive. They have an unpleasant odour, stain walls yellow and are unpopular. Some areas, e.g. sub-Saharan Africa, retain the use of DDT, believing that it is more effective and its risks are fewer than those posed by malaria. IRS, when used with appropriate insecticides and correctly and appropriately applied, can provide highly effective control of malaria. The use of IRS is one of the WHO's key tools against malaria and research is in place to provide safer and more effective insecticides.
Move communities further than mosquito flying distance from areas of standing water	To remove people from the effects of the mosquito and reduce its reservoir of blood	This can prove effective but there may be some problems. If people have to travel to a water source that is a breeding ground for mosquitoes, they may still become hosts for the parasites. Moving people from their traditional villages can cause conflict due to traditional links with the land and religious and social beliefs.
Mosquito traps	Act to attract, intercept and kill mosquitoes, reducing contact between the vector and humans	These take advantage of mosquitoes' sensory abilities which are usually used to find human or animal blood sources. Mosquitoes are attracted by odours, movement and heat. Once attracted, the mosquitoes are either trapped in containers or become stuck to a substance where they die. There are different types of trap. Some: • produce CO_2 to mimic the breath of an animal or human (these use simple methods of producing CO_2 often from a yeast–sugar mixture, to allow for durability and ease of maintenance locally) • emit **octenol** (naturally found in human sweat and breath, and given out by animals that eat vegetation), often used in conjunction with CO_2 • use flashing lights (in both visible and invisible wavelengths) at frequencies that alert mosquitoes to movement

▼ **Table 2.13** *(Continued)* Methods of approaching the control and eradication of malaria

Target/method	Reasons	Comment
		• emit heat to suggest a living creature to mosquitoes • use simple attractants such as sugar-based liquids (cheap, easy to manage and durable) to attract and drown the mosquitoes or sticky substances to hold onto the mosquito once attracted. Traps are placed in shaded areas between the source of the mosquitoes and where people gather, to prevent the mosquitoes from reaching areas where contact can be made with humans. This can be difficult in areas with a variety of, or unknown, sources of mosquitoes. Some traps, such as those that use light emission, can be difficult to maintain in areas with limited technological, electrical and battery resources. The simpler methods are much more viable in the long term in such areas. Traps can be very effective inside dwellings, where smaller areas are covered and the traps remain active at night. Recent studies show that effectiveness is good but depends on where the traps are placed. It is difficult to work out accurately the best places to locate traps as this often depends on local factors. To date, traps have been quite effective but only on a relatively small scale. Some newly developed varieties are predicted to be effective both indoor and out, providing protection for larger areas. One of these is the **attractive targeted sugar-bait trap**, which is expected to be in use by 2023. It is designed to protect areas where humans gather, e.g. agricultural areas. This trap is a large pad infused with fruit juice to attract mosquitoes and with targeted insecticides to kill them (but not other beneficial wildlife). Traps, on their own, are not a total solution, with effectiveness rated between 30% and 60%. They are effective as a tool in conjunction with other methods.
Mosquito nets ▲ **Figure 2.33** Treated mosquito net, Kibera shanty town, Kenya	To act as a physical barrier to mosquitoes; they are also usually sprayed with insecticides	These are a key tool in the WHO's strategy against malaria. They are relatively cheap, under £2 (approx. $2.50) each. They are very effective, particularly with the recent introduction of longer duration, double (two types of) insecticide-embedded varieties to reduce resistance. It is estimated that there would be around 30% fewer child deaths if children slept beneath treated nets. Basic versions of long-lasting insecticide-treated nets (LLINS) can last up to six years, but newer, even longer-lasting versions are being developed. It is estimated that over 150 million nets need to be supplied every year and this can lead to problems with transportation into remote areas.

▼ **Table 2.13** *(Continued)* Methods of approaching the control and eradication of malaria

Target/method	Reasons	Comment
Vaccines	Prevent people becoming infected (to create resistance to the parasite)	This could become the most effective manner of treating malaria if the treatment becomes cheap enough and easy to carry out. People would not be affected by malaria and the supply of infected blood would be reduced, leading to a reduction in infections and parasites. Many people believe that only an effective anti-malarial vaccine will bring about a large-scale change in the fight against the disease in endemic areas. Research into several vaccines is ongoing. The first licensed vaccine, RTS,S (Mosquirix), may be ready for full-scale use by around 2023. Its advantage is a predicted reduction in the incidence of malaria in children. Unfortunately, RTS,S is only for children and currently requires multiple doses. In its present form, it is only effective in 26% to 50% of cases and *only* against *Plasmodium falciparum*. Its durability is also not as long term as would be hoped. Progress on improving these limitations is being made. It is promising that there is increased development of vaccines with the possibility of availability within the next 10 years (for adults and children, and for protection against a range of *Plasmodium*). It should be noted that it costs over £500 million to take a vaccine from the laboratory stage to being a safe and effective product.
Genetic modification of mosquitoes	Decreases the number of mosquitoes and, in particular, females; lessens contact with humans; kills or disables the parasites, leading to reduced human infection and eventually eradication	Recent improvements in genetic research have led to gene drive techniques, which allow: 1 interference with survival or reproduction, causing: • male infertility, thereby reducing the population • a gender bias towards males, so reducing the female (malaria-spreading) population • infected female mosquitoes to die 2 alteration of targets for transmission by creating a *dislike* for humans as a target for feeding 3 neutralisation of the mosquito as a vector: • by modifying the female mosquito so that it is unable to carry the parasite • by interfering with parasite development so it dies or is unable to infect. Original efforts to make male mosquitoes sterile did help reduce mosquito populations but the fact that genetic engineering caused sterility meant that the gene did not spread. The gene drive approach has given a variety of tools, which can be used to attack a mosquito population while still allowing the alteration in the genetic code to be passed between generations and permitting the move towards eradication. Genetic modification relies heavily on scientific research/ techniques and production methods so therefore also relies on continued and large-scale funding for long-term success.

▼ **Table 2.13** *(Continued)* Methods of approaching the control and eradication of malaria

Target/method	Reasons	Comment
Prevention of spreading		
Insecticide spraying of aircraft	To prevent parasite-carrying insects from travelling from one location to another, and especially to stop drug-resistant strains from entering a region	As time and transport methods have improved, so has the potential for malaria to spread. As with all cases of malaria, it is important that quick diagnosis is made and action taken. Faster, long-distance (international) aircraft travel has added complications. Not only may a passenger have travelled long-distances before the symptoms appear, but mosquitoes can be transported as well. This has resulted in what is known as airport malaria. Not only have people on flights been bitten but the mosquito has survived for long enough to leave the airport and infect people in the local area where it has arrived.

Random searches at London's Gatwick Airport found that of 67 aircraft arriving from tropical countries, 12 contained mosquitoes. In temperate countries such as Scotland, warm, muggy summer **weather** may allow for longer survival of the mosquito. The UK, Australia, South Africa and other countries regularly spray aircraft (externally and internally) from areas where there are contagious or infectious diseases. Some countries also spray insecticides on and inside the aircraft while passengers are on board, e.g. India, Trinidad and Tobago and Uruguay. |
Early intervention and healthcare		
Provide local and community trained healthcare	Allows early detection and treatment, reduces long-term infection and may reduce numbers of parasite hosts for biting	The WHO highlights the need for early detection and containment. For countries without the financial ability to create a robust healthcare system, training locals in the basics of identification and early care can become an important step in the control of malaria. Funding or support for this must be maintained for it to have a long-standing effect.
Education ▲ **Figure 2.34** School malaria education in Malawi	To inform people of the basic steps to take to identify and treat malaria	With knowledge, people can take effective steps for prevention, identification and care. This assists in removing traditional responses and allows individuals to act to prevent malaria. Even simple advice such as how to use insect repellents, the use of mosquito nets and the need to have skin covered at active mosquito times (especially dawn/dusk) can be life-changing. Education on its own cannot halt malaria but it is an effective tool within a concerted effort including additional resources and aid. It is effective but will not stop every mosquito that bites.

Task

1   Describe the measures that have been taken to combat malaria and comment on how effective these measures have been.
2   Look at all the measures that have been taken (Table 2.13 on page 74) and identify those you believe would be the easiest and most appropriate for use by local communities. Explain your decisions.

Research opportunity

The terrible effects on people and economies caused by malaria have resulted in many charitable attempts to support the fight against the disease. In recent years, the Bill & Melinda Gates Foundation has funded research and aid programmes. You may find it helpful to research its attempts (or those of other charitable organisations) to find a solution to the problem. Try the following websites:

Bill & Melinda Gates Foundation – www.gatesfoundation.org/

Malaria No More – www.malarianomore.org.uk

Against Malaria Foundation – www.againstmalaria.com

Nets For Life – www.episcopalrelief.org/what-we-do/integrated-approach/malaria

Reflection

Let's pause for a moment to think about climate change and its possible effects on malaria. Here are a few points to consider:

●   Increasing global temperatures are already being blamed for observed changes in the locations of malaria, including it now being found at increased altitudes.
●   If global temperatures increase by just over 2°C, a further 60 million people will be exposed to malaria in Africa alone.
●   Malaria used to be endemic in the British Isles, with the last non-imported case reported in the 1950s. A mosquito capable of acting as a vector for the parasite still exists in the UK (*Anopheles atroparvus*). It is predicted that by between 2050 and 2075 temperatures and precipitation within the British Isles will be at a level that would support the parasite's reintroduction. Although numerous locations could become suitable for malarial infestation, the lower-lying areas around the Firth of Forth seem a likely candidate in Scotland.

What does the above information tell you about the climate conditions expected to be experienced around the Firth of Forth between 2050 and 2075?

Task

At this point in your studies, you should remind yourself of what you need to know for the exam. The SQA states that you should study:

●   a water-related disease: causes, impact, management.

If you have covered this topic fully, you should find the following exam-style questions quite straightforward.

Task

Causes

1 Explain the physical and human factors which put people at risk of contracting malaria.
2 Explain the importance of these factors in the cause of malaria.

Impact

3 Explain the impact on individuals, families and communities of endemic malaria.
4 Explain the impact on the economy of a developing world country of endemic malaria.
5 Explain the benefits to individuals, families and communities of the control or eradication of malaria.
6 Explain the benefits to a developing country of controlling malaria.

Management

7 a) Explain at least five measures that can be taken to combat malaria.
  b) Comment on the effectiveness of the methods you have selected.
8 Select three methods that you believe would be most appropriate for use by local communities to combat malaria. Give reasons for your answers.
9 Describe, in detail, the strategies used to manage malaria.

2.8 Primary healthcare (PHC)

This is a global approach to healthcare that aims to address people's health-related needs in the best possible way. Primary healthcare has at its core the belief that healthcare should be provided in a manner that is most appropriate to the circumstances of people and the area they live in. Needs, skills, experiences, expertise and understanding from local communities are viewed as being central to achieving the best system and outcomes.

It is understood that throughout the developing world many countries do not have the finances to create and maintain elaborate healthcare systems. In an attempt to find ways to improve health and expand access to even basic healthcare, developing countries have

adopted approaches that are more appropriate to their circumstances.

▲ **Figure 2.35** 'Barefoot doctors'

Much of what has become primary healthcare strategy around the world has been derived from the experience in China through what became known as the 'barefoot' doctors scheme. In the 1960s the communist government in China implemented a plan to provide basic healthcare for rural areas. The government realised that poor health conditions reduced the ability of the country to develop and produce food in quantities that would provide for the nation and create money through sales in international markets. Outside the main urban areas, there was very little access to healthcare facilities for millions of rural Chinese. With a lack of transportation and high levels of poverty, people were forced to walk for days to the nearest health centre. Even these centres were not always appropriately stocked or staffed. As a result, many people were not treated and diseases were able to spread quickly. Death rates and infections were high.

With a vast area of land to cover, and with limited finances, the Chinese Government designed a programme called the Rural Cooperative Medical System (RCMS) that took into consideration the need for localised medical care and the ability to provide it. Large numbers of medical professionals and 'volunteers' were dispatched to the countryside to work with, and for, the local people.

A key element of this plan was to train large numbers of local people in basic medical practices. These local medical auxiliaries then became the contact point for their community. They would deal with straightforward illnesses, assess patients and arrange for those who were more seriously ill to be referred to better equipped health centres or hospitals. This also assisted by allowing hospitals to concentrate on those with severe conditions. The auxiliaries were usually secondary school graduates who received an average of six months' training. This training focused on epidemic disease control, simple ailments and Western methodologies. They were not expected to deal with major surgical procedures but to provide simple effective medical treatment, advice on birth control, diet and food preparation and to dispense drugs such as anti-malarials. On some occasions, their knowledge on the transmission of diseases made them valuable assistants in the construction of water and sanitation systems.

The local community became responsible for the costs of running the service, but medical auxiliaries were not expected to work full-time; they remained as farmers or in other jobs in order to support themselves. They became known as 'barefoot doctors' – not full doctors and still very much part of the rural community.

In 1981 the barefoot doctor system was abolished, as a result of a change in economic policy, and within three years health coverage in rural areas had decreased by 85 per cent.

In 1985 the government announced that those original barefoot doctors could sit an examination and, if they passed, they would become 'village doctors', but if they failed they would be termed 'health workers'. Health workers would then work under the guidance of the village doctors. As such, the term 'barefoot doctor' was replaced.

Initially the system was very successful and rural health improved dramatically, but as time went on the system came under stress. Communities found it difficult to afford to maintain the local elements of the system such as the training of auxiliaries and the upkeep of health centres. It became impossible to train enough village doctors and health workers to deal with the high numbers of people who needed their assistance.

In an attempt to keep the system going, the government tried to encourage health workers and trained doctors to set up private practices. This resulted in two problems:

- Many doctors did not wish to leave the comfort of, and the potentially more financially beneficial, cities.

- The privately run practices often required the financially deprived rural workers to pay for treatment and, once again, this placed affordable healthcare out of their reach.

Although the Chinese Government attempted to put in place another programme, by 2003 the system was seen to have broken down, with evidence of an increase in TB, infant mortality and maternal deaths. Concerns were also raised at the possibility of increases in infectious diseases because immunisation rates had decreased.

In response, China increased its spending on health and introduced a government-run health insurance scheme. Payment rates to buy into the scheme were kept low, allowing individuals to purchase cover for treatment of serious diseases.

The legacy of the barefoot doctor scheme and its intentions to provide health coverage at a local and affordable level, dealing with the needs of people in their own area, is that the concept has been widely adopted elsewhere. Some schemes have run into similar difficulties, mostly due to a lack of money. Debt repayments and alterations in the world economic situation have made it difficult for the developing countries to maintain even their low level of financial input. Warfare, especially across the continent of Africa, has created no-go zones and totally disrupted health initiatives.

In the 1980s the Nicaraguan Government ran a programme based on similar concepts, building 500 new health clinics within local communities. Local people were encouraged to take part and received basic medical care and health education. This project saw a drop in infant mortality from 30 per cent to 8 per cent, mass vaccination eradicated polio and reduced the incidence of whooping cough, and malarial infections dropped significantly. Unfortunately the scheme's effectiveness was reduced by war, a reduction in the world economy and natural disasters.

In general the intentions and benefits of such schemes outweigh the disadvantages. They deal specifically with local people's needs and can break down barriers by involving known and trusted members of the community. They are most effective in rural areas that are often remote and isolated, with limited existing educational and medical support.

Another important element is the concentration on low-tech, low-cost solutions that can be maintained by the local community and reduce reliance on expensive foreign medicines or aid.

Task

1 What were the conditions in China that caused the government to set up the Rural Cooperative Medical System (RCMS) that would become known as the 'barefoot doctor' programme?
2 Describe the way in which the Rural Cooperative Medical System worked.
3 Why were the medical auxiliaries who were trained known as 'barefoot doctors'?
4 Why do you think it was important that the barefoot doctors came from, and lived in, their villages or communities?
5 Evaluate the success of the barefoot doctor scheme and give reasons for its successes and failures.

In Nicaragua there were attempts to reinvigorate the approach to primary healthcare. In 2014, Dr Michael Mangold sought support for an approach based on the Chinese barefoot doctor scheme, but also making use of advancements in technology. Dr Mangold's model mirrored the Chinese one in many ways:

● The community 'barefoot doctors' would come from the country itself and, if possible, from the local community area.
● The people to be served would decide whether to accept the doctor or not, and if accepted the doctor would live within the community. The Chinese experience was that if the people felt they had ownership of the process a greater level of trust developed, increasing the effectiveness of the service.
● The barefoot doctor would not be utilised only for medical interventions but as an educator, giving advice and information to improve the general health of the community (for example, nutrition, clean water and sanitation).

The proposals also suggested that modern barefoot doctors should be trained in computer literacy, including the use of mobile phones and apps. After six months of classroom training and another six months working in the field, the barefoot doctors would qualify. Not only would their basic training be put into practice,

but the barefoot doctor would be supported by the use of mobile phones or similar devices. These would have an internet connection and be supplied with apps giving access to up-to-date techniques that would be suitable for use. Similar apps were already in use in developed countries to assist doctors and surgeons. Dr Mangold realised that this could be costly but suggested that charities, such as those that exist in the UK and USA which collect and refurbish phones, could be seen as a source for equipment.

The doctor visualised a time when access to artificial intelligence to assist with surgery and other complex medical issues, such as taking and interpreting electrocardiograms, would be the norm. Unfortunately, the doctor's ideas did not reach completion in Nicaragua as he was forced to leave when he was injured in an accident.

As technology and access to satellite communications improve, iPhoot doctors (also known as 'smartfoot' doctors) may form an important step forward in primary healthcare. Technology could bring about major breakthroughs in worldwide healthcare.

Mobile devices are regularly used in the developed world within healthcare systems, and specialist medical apps are available. This technology could become core to the delivery of effective healthcare within developing countries by allowing real-time, interactive access to medical records and health information, and to doctors and other healthcare practitioners where there is little or no specialist care.

In Zambia, the use of mobile and satellite phones has become an integral part of attempts to eradicate malaria, with communication allowing fast reaction to outbreaks, contact for assistance, relay of information, monitoring and access to real-time specialist advice. In addition, drones with cameras are used to locate mosquito breeding areas, create up-to-date maps of infected areas, spray infected areas and mosquito breeding areas with insecticides, and transport materials and drugs. Although initially operated by specialists, locals are being trained in the operation of these systems. Some people are concerned that, although locals are trained in the operation of such technology, without continued funding and regular replacement of technology, the scheme will be time limited. In such a way, this initiative may not be truly self-sustaining in the manner expected from primary healthcare.

Research opportunity

Do you think that Dr Mangold's ideas for artificial intelligence, mobile phone and tablet use to assist local communities in the developing world are workable ideas in the present? Explain your reasoning. Try doing an internet search using these suggestions:

- medical apps   ● apps for doctors
- do-it-yourself electrocardiograms
- apps for surgeons.

Do you think that Dr Mangold's proposals were a good idea? What is your reasoning behind your decision?

WHO and primary care

In 1978 the WHO took steps to find international agreement and formalise an approach to, and the definition of, primary healthcare. The intentions and some of the methodology behind the barefoot doctor scheme were openly stated as the inspiration for this new approach to healthcare worldwide; not just for developing world countries but for developed countries too. The International Conference on Primary Health Care in Alma-Ata (now known as Almaty), Kazakhstan in 1978 agreed a declaration that would become central to the WHO's policy of 'Health for All'. The Declaration of Alma-Ata defines primary healthcare as:

> '... essential health care based on practical, scientifically sound and socially acceptable methods and technology universally accessible to individuals and families in the community through their full participation and at a cost that the community and country can afford ... in the spirit of self-reliance and self-determination.'

Source WHO
www.who.int/publications/almaata_declaration_en.pdf?ua=1

The approach is one that demanded a massive change in the way that health provision was seen and organised throughout the world. The focus would be on the people themselves, those who needed the medical treatment, and finding ways that would allow appropriate medical care to be provided. It required that government policymakers take into consideration the views of individuals' local circumstances and agree to embed the priorities of health in all policies and government departmental decisions. The advantages of good health and quality of life to the country's economy needed to be highlighted to all those involved in government.

In simple terms, the new approach called for:

- the participation of local communities in deciding healthcare priorities
- the focus of healthcare responding to the needs of the community and emphasising preventative medicine and methods
- links between healthcare and trade, industry, economics, politics and social issues to be established.

The policy demands that all partners in the process work together to satisfy the core aims of primary healthcare. Attempts need to be made to fashion policies that fit closely with the ability to maintain access to healthcare that is sustainable. In 2009, the Director of the WHO, Dr Margaret Chan, identified previous failings in earlier approaches to improvements of healthcare when she stated:

> '... developing countries are littered with the debris of poorly co-ordinated aid, with dilapidated clinics at one extreme, and un-used hospital beds at the other.'

Dr Chan also stated the need for control of health plans to be 'country owned and country led' and not dictated by donors who may only have had short-term input in mind. She also stated the need for countries to take responsibility and see the value of improved healthcare. African countries pledged 15 per cent of their annual budget to go to healthcare and Dr Chan encouraged them not to renege on this as improving national health would have a long-term benefit in terms of raising finances and improvements to GDP.

The WHO report on primary healthcare highlighted four key points that must be carried out to ensure that the intentions succeed:

1 Governments must set up systems of accountability and include the voices of experience from all of their society in planning and accountability systems.
2 Health inequalities must be tackled through universal coverage of health facilities. Governments must demonstrate the results on the ground and show that a fair distribution of resources is in place.
3 Healthcare should be positioned centrally in all policies and not just health. Governments must share and encourage the view that health gains have potential benefits for all sections of society and government.
4 People need to be put at the centre of healthcare – their expectations and their experiences.

 Task

1 Outline the basic principles behind the WHO's primary healthcare policy 'Health for All'.
2 Why is it important for local communities to participate in deciding healthcare priorities?
3 The WHO believes that 'Healthcare should be positioned centrally in all policies and not just health.' What are the advantages to a developing world nation of doing this?

So far, the WHO approach to primary healthcare has had mixed results. Where put into practice, there is a general improvement in healthcare, but the pace varies from country to country and across the continent of Africa it is particularly slow. In African countries, infant and child mortality is only slowly improving while in some areas maternal mortality rates are getting worse.

Although being seen as a great improvement in healthcare worldwide, similar problems to previous attempts are beginning to arise. The WHO has identified a lack of commitment by some countries to push through the changes that are needed to fulfil the principles of primary healthcare. Across the continent of Africa in particular, weak structures, poor operating systems, a lack of rigorous monitoring and evaluation and inadequate attention to the principles

of primary healthcare have found the continent lagging behind other areas of the world. Much of this has been put down to economic conditions creating a reduction in GDP and so a reduction in the money available. In addition, a rising number of conflicts have seen funds being transferred to military resourcing and some areas have again become difficult to access.

Although many of the problems are seen to be the same as with other attempts to provide healthcare to the most economically deprived areas, the advances made at present show that this is a great opportunity to create a leap forward in healthcare for all.

Recent approaches to healthcare in Africa

Even with the noted disappointments in the process towards better provision of healthcare across the continent of Africa through a primary healthcare model, the improved ability of the WHO to identify weaknesses and failures, as well as to highlight benefits of the successful models, has reinvigorated moves towards change.

The continent of Africa faces multiple health challenges on a large scale. Low levels of development already make dealing with these challenges extremely difficult and the complexity of these continues to evolve. The traditional fight against tropical diseases, diarrhoea, maternal and child mortality has been added to by an increase in HIV/AIDS cases, while populations and life expectancy are growing.

There is an ever-growing realisation that the structure of many healthcare systems in the continent is focused on treatment of the results of poor health and not a reduction of the causes. As the traditional battles are added to and the population grows, these systems become massively overstretched and unfit for purpose.

Dr Ernest Darkoh, a specialist in African healthcare, strategic planning and systems development, is clear that just building more hospitals or clinics is a sign of failure and states: 'We must make disease unacceptable instead of building ever larger infrastructure to accommodate it.' His view fits squarely within the beliefs of the WHO that the reduction of the causes of disease is of much more benefit than continually attempting to treat its results. A decrease in the stresses on the treatment of a high volume of the traditional diseases could allow for reductions in costs resulting in increased quality of acute healthcare and finances available for wider health provision.

There is an awareness that there needs to be a development of models of care which take into consideration the basic situations within rural and remote areas and their own specific needs. This requires a significant overhaul in the mindset, structure and human resources in the healthcare systems across the continent of Africa that are generally modelled on those of more developed nations, which have greater financial resources and very different circumstances.

The outline being suggested and forming the most up-to-date strategies across the continent includes the following:

1 Hospital and clinic-based models where people must come for healthcare are being replaced by healthcare that comes to and interacts with people.
2 Health information and education is being provided throughout society.
3 'Task shifting' within the health system, which allows non-medical aspects to be taken away from trained medical specialists (doctors and nurses), freeing them to focus on acute care.
4 In line with above, local people are being trained to educate their community on health and disease prevention. This reduces costs and helps build a model of, and the belief in, individual, family and community ownership of health.

There has also been an awareness of the difficulties of maintaining healthcare structures due to low levels of income for the system and a reliance on international aid. Many countries are seeking methods that will allow sustainable financing. One of the major aspects that must be taken into consideration is the ability of people to afford medical care. If people are forced to pay to access healthcare, they may simply not be able to afford it. This leads to a reduction of usage which, in turn, may lead to a continuation of the core problems of poor health. Ghana and South Africa were two of a number of African countries that abolished primary healthcare fees in the public sector in acknowledgement of this.

Already some insurance companies are looking at plans to introduce small-scale private insurance plans in Tanzania, Ghana and Malawi for those who do not have access to employment-related or other private schemes. These schemes would still not apply to the most financially deprived in society and support is still required through other methods. Ghana has increased taxes on products for sale within the country to pay for a social insurance health insurance that benefits all members of society.

By 2018 it was deemed necessary to stress continued support for the ideals of the Declaration of Alma-Ata, made some 40 years earlier. The Global Conference on Primary Health Care, Astana, Kazakhstan (2018), stated that primary healthcare had made great strides globally in improving access to healthcare and finding effective ways towards sustainable health systems for all.

But there have been large-scale changes since 1978. The world's population has more than doubled; although more people are living longer this has added an extra focus on dealing with illnesses related to older age. It is also acknowledged that over half of the world's population still lacks access to essential life-saving healthcare.

The Astana conference found agreement that a primary healthcare approach is still the most effective one for solving healthcare problems.

Primary healthcare in Ethiopia

Ethiopia can give an almost unique insight into the development of a system with primary healthcare at its core. Following decades of extreme famine and civil war, the country had virtually no healthcare strategy or provision except for missionary clinics and international donor-run hospitals. The majority of the country's population relied on traditional and spiritual healers. In the last 20 years this situation has changed dramatically, with more than 85 per cent of the population now having access to primary healthcare and infant mortality declining by 52 per cent to 41 per 1000 live births.

Initially the government decided on a top-down structure where it made the decisions for the whole country and was criticised for not fully engaging with those at the local level or who had experience there. The government's viewpoint was that it needed some immediate, emergency action to get some kind of system in place as there was nothing there in the first place but a great need for healthcare. In the last few years, there has been a movement towards broadening and deepening engagement with local communities and truly creating a primary healthcare system in line with what is generally expected through the WHO guidelines.

The government is now targeting young women to fulfil the roles of 'health extension workers' (trained in education, prevention and low level treatment), in the belief that they are the closest to those who are the main beneficiaries of primary care through their traditional role in Ethiopian society. Women's groups will also be linked into the system to provide a channel for local concerns and ideas.

The healthcare structure being developed by Ethiopia has three main strands:

1 Primary healthcare delivery including low-cost approaches (40,000 trained and deployed health extension workers).
2 The building of health centres for more complex medical treatment to work in parallel with primary healthcare.
3 Private-sector investment in new hospitals.

The structure above has been put in place to fit with the particular needs of Ethiopia developing a complete healthcare structure from almost zero. In other African countries the problems may come from having to restructure an existing system.

Primary healthcare in Cuba

Although Ethiopia and many other African nations rely on a mixture of government, local and private elements within their systems, across the Atlantic Ocean, Cuba has taken another approach to primary healthcare. Here primary healthcare is placed directly in the centre of the health system as a point of law and policy. All health systems are run through primary healthcare. Great importance is put on local communities identifying healthcare problems and priorities then working with government representatives to implement a community-centred plan.

Since this move to a focused primary healthcare strategy in the 1980s, healthcare indicators have improved significantly. Cuba's efforts have resulted

in some similar population indicators to those in developed countries such as those in Table 2.14.

▼ **Table 2.14** Cuba and UK comparison

Indicators	Cuba	UK
Infant mortality rate per 1000 live births	4.2	3.7
Life expectancy (years)	79.9	81.7

 Task

You now need to remind yourself again of what you should know for the exam. The SQA states that you should have covered:

- primary healthcare strategies.

If you have covered this topic fully, you should find the following exam-style questions quite straightforward.

1 a) Explain what is meant by primary healthcare.
   b) Evaluate why primary healthcare is seen as an appropriate strategy for developing nations.
2 a) Explain some specific primary healthcare strategies.
   b) Evaluate why these strategies are particularly suitable for developing countries.
3 a) Give examples of primary healthcare in developing world countries.
   b) By referring to specific examples, identify the strengths, weaknesses and problems experienced.

Summary

This chapter has explored development and health, two separate but linked topics. The many dimensions of development have been identified, defined and measured. Differences exist between and within countries of the world. Key indicators of development are population health and the incidence of disease. This chapter has identified factors influencing health and studied in depth the causes, impacts and management of malaria, as well as the impacts of primary healthcare.

Introduction to Global Climate Change

▲ **Figure 3.1** Oil refinery

The aim of this chapter is to help you to develop and apply your knowledge and understanding of climate change. This is a massive global issue, with its threat to our living conditions and even the continued existence of humans on the planet. With such an important topic, it is essential that we all become well informed. This chapter should be seen as a springboard for your own investigations, encouraging you to access the most up-to-date information and not just being bound by what you read here.

For the exam, you need to be able to demonstrate the interaction of physical and human factors in creating climate change, and evaluate strategies adopted in the management of this issue. An appreciation of sustainable development to reduce negative effects of climate change is important.

The key areas are:

● physical and human causes
● local and global effects
● management strategies and their limitations.

A better understanding of climate change will be achieved if you already have an awareness of how the

Earth's atmosphere works, especially in relationship to the global heat budget and energy redistribution. When taking into consideration the natural drivers of climate change, a strong knowledge of processes in the lithosphere is also helpful.

3.1 Climate

Climate is the *average* pattern of the elements of weather prevailing in an area over a long period of time (around 30 years or more). This average *suggests* what *could* be expected at a location or region during a particular time of year but it does not state that it *will* happen. Climate should not be confused with weather, which is only the atmospheric conditions (as experienced within the troposphere) at a particular location and at a specific time.

A region's climate is generated by a number of factors in a complicated interaction. Without an atmosphere there would be no weather and no climate zones, just a dry surface being directly radiated by solar winds and short-wave solar radiation from the Sun. However, with the atmosphere in place, the system creating the climate is powered by the incoming solar radiation (insolation). This creates the conditions which lead to global energy transfer, stimulation of atmospheric and oceanic circulation, the movement and changes of state of water into vapour and the creation of clouds and precipitation, pressure belts and wind patterns. Land shape and altitude also have influence, with mountains deflecting or forcing winds to rise and cool, creating areas of precipitation and rain shadow. Surface types and even vegetation can change the albedo of an area, or influence the amounts of water and vapour in the surrounding area. These are only a few of the influencing factors in what is an extremely complex and finely balanced system where the atmosphere, hydrosphere, cryosphere, lithosphere and biosphere all add elements to climate creation.

Climate is not a static entity. As with any system that relies on so many variables and influences, alterations to any of these contributing factors can result in change. The history of planet Earth is one of development to its present state, with atmospheric

▲ **Figure 3.2** Polar

▲ **Figure 3.3** Tropical rainforest

▲ **Figure 3.4**
Tropical desert

change, plate tectonics and even changing amounts of insolation being some of the alterations to the system that creates climate. Scientific research confirms that the Earth's climate has been different over numerous timescales and the system maintains the ability to modify and react to changes in the conditions which create it.

3.2 Climate change

It is important that we have a clear understanding of what we actually mean by climate change. For our purposes we need to accept that there are numerous physical contributors to climate change that influence climatic development (and have done even during the vast length of time that humans did not exist on the planet). We must also be aware that human beings have altered many of the elements that contribute to the climate creation system and that anthropogenic (human) influence is active.

The simplest way of defining climate change is as a statistically proven (evidence-based) alteration of the climate system over an extended period of time, regardless of what has caused its modification. In addition to this we need to recognise that it may also include a redistribution of the frequency or intensities of weather events around the average conditions. An example of this would be increased storm activity with a higher intensity of precipitation over short periods but with fewer occurrences of regular but moderate rainfall. Here the average conditions would stay the same but the different patterns of weather could change the climate zone considerably by favouring different flora and fauna, changing erosional patterns and amounts, or altering traditional agricultural methods, for example. Not all of the alterations may be negative and in some areas humans may find the shift in patterns beneficial.

Changes in climate may happen over a variety of timescales. Changes in the chemical make-up of the atmosphere, and the patterns of heat distribution through the atmosphere and oceans, along with changes in surface vegetation, have the potential to alter climate within a relatively short timescale. Plate tectonic movement is a much slower process, moving the positions of continents, surface types and even mountains. This movement can block the path of ocean currents, change a region's albedo and create mountains that block or alter wind patterns. All of these can destabilise and drastically modify the patterns of atmospheric motion, heating and the transfer of energy around the planet. These are just examples and more investigation would highlight a much larger assortment of potential drivers of climate change over every possible timescale.

Sources of evidence for climate change

There is a general scientific agreement that our planet experiences climate change and has done so on numerous occasions in the past. This consensus is based on evidence collated from a variety of research methods. No single method's results would have encouraged scientists to come to such an agreement but, due to the diversity and amount of corroborative data, climate change is seen as a continuing and active process on Earth.

Since the nineteenth century, weather and climate statistics have been collected around the world. These have become more accurate and reliable over time.

In the twentieth century, a mostly land-based global network of weather stations provided climatologists with an extensive statistical record, although it still left large areas of ocean and the most remote areas without coverage. The introduction of satellite technology in the 1970s achieved global monitoring and global vegetation coverage was also identified and quantified. This enables comparisons to be made on a regular basis and large-scale changes identified.

Although there is a great deal of scientifically recorded detail about recent climatic conditions, much of what climatologists and palaeoclimatologists have learned about ancient climates has come from indirect evidence (climate proxies) through which the state of the climate can be inferred. Table 3.1 shows the techniques that are used to reconstruct past climates, collect data and gain evidence for climate change.

Like all methods using proxies, there is the possibility of error. To help lower the chance of this, more than one methodology is used and comparisons are made to confirm the results as much as possible.

▼ **Table 3.1** Methods used to provide climate change evidence

Palaeoclimatology	Archaeological/historical	Modern scientific data collection
Palaeoclimatologists (those who study **palaeoclimatology**) use a wide range of **climate proxies** to work out the climate and atmospheric conditions of the planet in times before recorded human history. Analysis of the chemical make-up and properties of rocks, sediments, ice sheets, shells, corals and fossils shows the conditions when they were formed. Using scientific understanding of the conditions needed for the existence of particular flora and fauna, their fossils or remains can also infer the climate of the time: • Rock, sediment, coral and fossil analysis • **Speleothem** analysis (e.g. **stalagmites** and **stalactites**, analysis of water trapped in layers) • Ice core, lake and ocean bed sediments analysis • Carbon dioxide levels within trapped air bubbles (amounts of carbon dioxide and different isotope levels) • **Pollen analysis** • Oxygen isotope analysis (from trapped air bubbles) • Dendroclimatology and **dendrochronology** • **Radiocarbon dating** • Environmental change (through, e.g. fossil records of plant and animal change or extinction) • Sea-level change (deposition and archaeological evidence)	• Analysis of human/animal/crop/ food remains • Artefact analysis (especially those made from organic materials) • Cave paintings/drawings (showing local conditions and even animals, plants and general environments) • Inscriptions (on monuments, ancient buildings and even the caves of primitive dwellers have inscriptions and carvings) • Early rainfall records (e.g. Greece and India) • Written descriptions (e.g. travellers' notes, diaries, agricultural records, early weather data collection and reports)	Evidence is collected to measure all the elements of weather and climate (including sea/ocean levels, sea/ocean temperatures and atmospheric gas levels) in a scientific and co-ordinated manner through: • Satellites • Worldwide network of land-based weather stations • Aircraft • **Radiosonde balloons** (can take measuring instruments up to very high levels of the atmosphere) • Ships • Buoys (used to collect a variety of oceanic information at different sea/ocean levels. Can also be used to track currents and report on their conditions) • Computerised collection, storage, analysis and electronic sharing of data • Ice mass, thickness, movement, extent measurements In addition: Observation of environment change through life-form activities (alteration in patterns of distribution and adaptation)

With the increase of scientific knowledge and technology, those involved in the study of climate change have been able to carry out more and more detailed research, create more accurate models and, with the use of computers, make complex calculations and correlate with other gathered information.

Research opportunity

Although not specifically needed for the higher exam, you may find it interesting to look into the methods used by palaeoclimatologists, historians, archaeologists and climatologists to collect data about climate change as shown in Table 3.1.

What has research shown?

- Climate conditions are not a constant. Climate constantly alters to reflect prevalent conditions over short, medium and long timescales.
- Climate change does exist and has occurred regularly throughout the Earth's history (approximately 4.6 billion years).
- A variety of climate zones may be found at the same time around the planet but they are created and influenced by the overall planetary conditions in existence during that period.
- Temperatures have ranged from extreme heat to extreme cold.

▲ **Figure 3.5** Prehistoric rock art – Akakus, Sahara, Libya

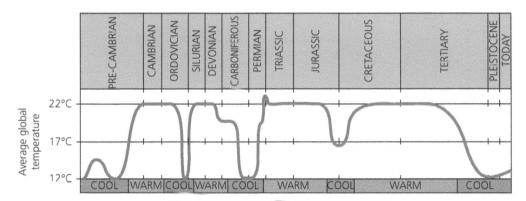

Pleistocene 2.59 million to 11,700 years ago
Tertiary 66 million to 2.59 million years ago
Cretaceous 145 million years ago to 66 million years ago
Jurassic 200 to 145 million years ago
Triassic 252 to 200 million years ago
Permian 299 to 252 million years ago
Carboniferous 359 to 299 million years ago
Devonian 417 to 359 million years ago
Silurian 443 to 417 million years ago
Ordovician 485 to 443 million years ago
Cambrian 542 to 485 million years ago
Pre-Cambrian 4.6 billion to 542 million years ago

▲ **Figure 3.6** Earth's long-term temperature variations

- For the last 2.59 million years (**Quaternary Period**), Earth has experienced global temperatures that have fluctuated from extreme cold to relative warmth. This has resulted in a cycle of glaciation lasting for around 100,000 years followed by **interglacials** of approximately 10,000 years. When glaciation was at its height, continental ice sheets reached as far as 40° latitude. The last glacial period ended between 15,000 and 20,000 years before present (BP), although around 13,000 years BP there was a short period of roughly 1000 years which saw a return to colder conditions. This was followed by a period of rapid warming. This pattern suggests that we are within an interglacial period at present.

- In the last 10,000 years temperatures have been relatively warm by Quaternary standards, although there have been fluctuations in conditions. Warmer periods were found during the Roman and then medieval times, with a much cooler period from around 1300 to 1870, which is referred to as the **Little Ice Age**. Although not really an **ice age**, this much cooler interval was at its height between 1600 and 1800.

- Throughout the twentieth century, and up to the present date, there has been an upward trend in global temperature. Atmospheric temperature has increased by above 0.8°C and oceanic temperature by above 0.1°C. Much of this has been since the 1970s.

- Since the 1970s the upward trend in global temperatures has accelerated to a rate never before experienced in the time of human existence (modern humans are believed to have evolved some 200,000 years ago).

- By 2019, 18 out of 19 of the warmest years ever recorded took place in the twenty-first century.

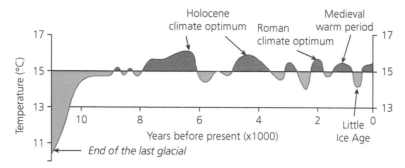

▲ **Figure 3.7** Temperatures since the end of the last ice age

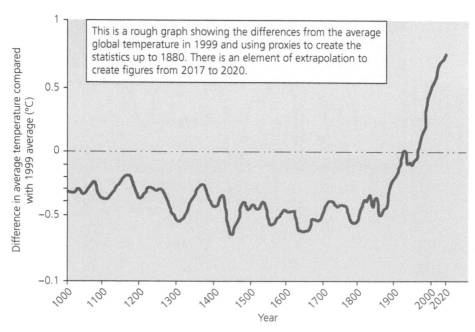

▲ **Figure 3.8** Difference in average temperature compared with 1999 average

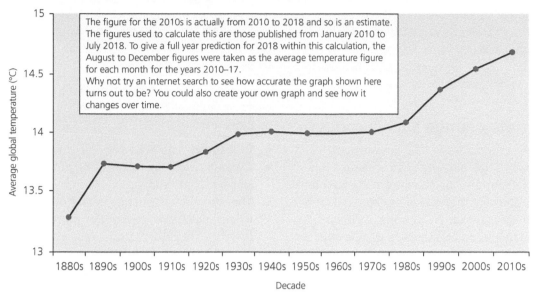

The figure for the 2010s is actually from 2010 to 2018 and so is an estimate. The figures used to calculate this are those published from January 2010 to July 2018. To give a full year prediction for 2018 within this calculation, the August to December figures were taken as the average temperature figure for each month for the years 2010–17.
Why not try an internet search to see how accurate the graph shown here turns out to be? You could also create your own graph and see how it changes over time.

▲ **Figure 3.9** Temperature change, 1880s to 2010s

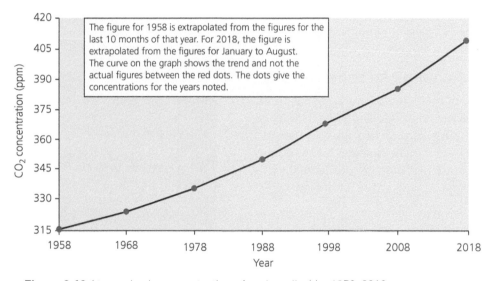

The figure for 1958 is extrapolated from the figures for the last 10 months of that year. For 2018, the figure is extrapolated from the figures for January to August. The curve on the graph shows the trend and not the actual figures between the red dots. The dots give the concentrations for the years noted.

▲ **Figure 3.10** Atmospheric concentration of carbon dioxide, 1958–2018

- July 2018 marked the 402nd month in a row where the temperature across the world's land and ocean surfaces was higher than the average temperature from the twentieth century (twentieth-century average was 15.8°C). This was the fourth highest temperature for July since records began in 1880.
- Five out of the world's six continents had a July 2018 temperature that was among the nine highest recorded since 1910.
- July 2018 was Europe and Africa's second highest temperature ever recorded for that month.
- The incidence of heatwaves across the globe has increased in recent years, while cold snaps are shorter and milder. In general, the number and severity of extreme weather events has increased.
- Both snow and ice cover on the planet are decreasing, as is the mass of ice in polar regions.
- Large areas of former permafrost have begun to melt.
- A variety of animal and insect species have been observed moving from their traditional locations to ones at higher altitudes or different latitudes as a response to changing environmental conditions.
- In recent times, the amount of carbon dioxide in our atmosphere and oceans has vastly increased. During 2013 in the atmosphere alone, the

concentration of carbon dioxide (CO_2) exceeded 400 ppm for a sustained period for probably the first time in at least 800,000 years and possibly even the last 4.5 million years. The increase has continued and in May 2019 it was recorded as 415.26 ppm.

Why is climate change such a concern?

It seems that everyone agrees that climate does change, so why has this become such an issue? It is not so much that climate has changed in the past that is the worry for the vast majority of scientists; it is the speed at which it appears to be altering at present. Added to this is the recognition that human beings have never had to experience such a rapid alteration in conditions and that we are unsure what these changes will bring. We are left wondering what effect the increasing temperatures and additional gas content will have on our climate. Will humans be able to adapt to the changes or will we simply be another species that disappears due to environmental changes?

Why is climate change so controversial?

Our understanding of the complex systems that create and alter climate is not complete and our scientific methods are only slowly coming to grips with the subject. Following recent increased and higher quality research (and the more accurate data collected), scientists have come to a consensus identifying:

- the actions of human beings as being the major driver of the recent increases in global temperature
- anthropogenic activities (human, industrial, agricultural and social activities) as having altered the conditions of the planet, leading to temperature increase
- the burning of fossil fuels as a massive influence (especially through industrial processes) by adding vast amounts of gases to the atmosphere that encourage increasing temperatures
- increasing temperatures as stimulating climate change
- the rapid rise in planetary temperature as stimulating change to the climate system, resulting in an overall negative effect on the lives of human beings and other life forms.

Much of the controversy around climate change is stirred by a lack of proper understanding by some people who comment on the subject, including leading world politicians, or misinformation spread by those with vested interests who would not wish their products to be taken off the market if they are shown to cause climate problems.

The term global warming has become central to the discussion about climate change. Be careful how you use this term:

- Global warming is powering climate change.
- Climate change is the altering of the pattern of world climate zones.
- Through climate change, some places may experience more frequent or more severe instances of colder conditions, while others may have more frequent or more severe instances of warmer conditions.

The term 'global warming' is more correctly used to describe the recent planet-wide increase in temperature believed to be created by human activity through global industrialisation (approximately 1800 to present).

Task

1 Explain the difference between the following terms:
   a) climate and weather
   b) climate change and global warming.
2 With reference to Table 3.1 (page 90), create a spider diagram or a mind map to show the sources of evidence for climate change.
3 Describe the changes in climate:
   a) throughout the Quaternary Period up until 10,000 years ago
   b) from 1900 to the present day.
4 Why is there concern about recent climate change?

3.3 Causes of climate change

There are many different influences that shape and change climate. These are known as climate drivers or climate forcing mechanisms. Climate has physical (natural) drivers that interact to create and influence climatic conditions. Climatologists now also identify the activities of human beings (anthropogenic) as being an additional driver, engaging with/altering the natural dynamics of the climate system.

Due to the complex and interrelated nature of the climate system, each of the climate drivers does not result in a simple or predictable climate change event or effect. The climate in different eras may present different conditions for the drivers to act on. This results in varying levels of change, reactions or outcomes compared with another period which exhibits initially dissimilar climate conditions.

While climate drivers act from outside to add to or change the climate systems, they also stimulate feedbacks within it. **Climate feedback** is the process through which changing one element has a knock-on effect that alters another. It can even have a domino effect, causing multiple variations in the system. Positive feedback amplifies a change whereas negative feedback reduces it.

Physical or anthropogenic drivers both create **feedback effects**. For example:

- Additional greenhouse gases can trigger increased heating in the atmosphere, which warms the seas and oceans, causing more water vapour to be released. This intensifies the **greenhouse effect** and a self-reinforcing cycle is put in motion (Figure 3.11).
- Expanding areas of snow or ice cover create an increase in the reflective quality of the surface and so stimulate further cooling. This encourages more snow and ice cover to form, further amplifying the effect.

These are only some examples; alterations to gas balance, cloud cover and even ecosystems can put in motion feedback cycles.

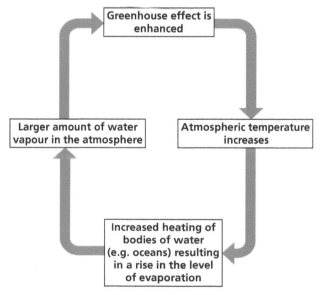

▲ **Figure 3.11** Positive feedback cycle driven by additional greenhouse gases

Physical drivers of climate change (natural forcings)

Energy emission by the Sun

The vast amount of energy affecting and driving the Earth's climate system comes from the Sun. The Sun is mostly made up of a hot plasma of hydrogen and helium gases and creates energy by thermonuclear fusion at its core. This energy is then radiated out and into the solar system as heat and light. Earth only intercepts a very small amount of this energy.

Earth receives its energy as incoming short-wave solar radiation from the Sun. This is also known as insolation (a good way to remember this is by thinking of it as *IN*coming *SOL*ar radi*ATION*). This moves downward through the atmosphere interacting with each of its layers, reducing the final energy amount that reaches the surface. Some of the energy is reflected into space. Although there is atmospheric heating from the incoming energy, much of the warming comes from the energy being returned from the Earth's surface (long-wave terrestrial radiation) and through latent heat given out when water evaporates, rises *from* the surface and condenses.

The global heat budget is the usable energy maintained by the Earth as a balance between energy received by the planet (input) and that which is radiated back out into space (output). This balance of energy powers the atmospheric system, ocean currents and climates which allow life to be sustained within the biosphere.

The rate of emission by the Sun can vary over time. Although in a relatively stable condition, the Sun has periods when it emits additional energy and others when it appears less active.

Extra activity on the Sun can result in huge magnetic storms called **sunspots** that appear on its surface as darker patches (Figure 3.12 on page 96). The frequency and intensity of sunspot activity varies around an 11-year cycle. Periods of high sunspot activity result in increased brightness and more energy being released by the Sun and received by the Earth. Change in the Sun's brightness, and its emission of energy, especially over longer periods, is believed to have an influence on the climate.

At present, average sunspot activity is in decline and when compared to evidence from ice **core samples** it appears to be happening at a much faster rate than

▲ **Figure 3.12** Sunspots

has been seen in the last 9000 years. This should result in a reduction in global temperatures. This is not happening and raises three possibilities:

1   Other climatic drivers are influencing the process and countering the effect of sunspots.
2   Sunspots do not have as great an effect on global climate change as previously believed.
3   Sunspots have a more complex relationship with climate change, which may also explain the movement in regional effects observed.

Since the 1970s scientists have been able to monitor solar output much more accurately and, by using satellites, without the interference of the atmosphere. Solar output has remained relatively constant and has not shown a net increase in the last 45 years and so cannot be responsible for the global warming observed.

By using sunspot records, the level of solar activity can be estimated for a longer period. These records suggest that in the twentieth century solar activity increased until the early 1950s but has decreased since. This evidence allows some to infer that solar output could have influenced global temperature rises up to the 1950s, but the decrease since then should have signalled a cooling. This has not happened, and temperatures continue to rise.

If solar activity was driving global warming, it would be expected that a uniform temperature increase would be found in each level of the atmosphere. Radiosonde balloons and satellites have provided extensive data which suggest that although the lowest layer of the atmosphere (the troposphere) is experiencing warming, the **stratosphere** is in fact cooling. Climatologists believe that this is exactly what would happen due to an increased amount of greenhouse gases being present in the troposphere, creating an **enhanced greenhouse effect** (see page 112).

Variations in the Earth's orbit and movement

There are three types of change to the Earth's orbit and movement that affect climate change:

- eccentricity (the stretching orbit)
- axial precession (wobble)
- axial obliquity (axial tilt).

Milutin Milanković identified repeating cycles of these as reasons for the advance and retreat of polar ice over the millennia and created mathematical models to show the variations. The movements described by Milanković suggest reasons for changes in the amount, timing and seasonal receipt of solar energy. The cycles have been named after him and are known as **Milankovitch cycles**.

Eccentricity (the stretching orbit)

The Earth's orbit around the Sun (a journey that takes one year) moves from being almost circular to elliptical (an oval-shaped pathway) over 96,000 to 136,000 years (Figure 3.13). A 'circular' orbit means that the planet's distance from the Sun is the same all year round. As such, seasonal variations of energy receipt and temperatures are minimised. However, as the orbit

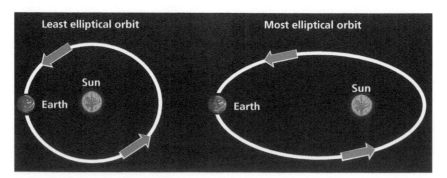

▲ **Figure 3.13** Earth's eccentricity

becomes more and more elliptical, seasonal variations become extreme, as do temperatures.

At present the shape of the Earth's orbit is only very slightly elliptical with only around 3 per cent difference from its farthest point (aphelion) compared to its closest point (perihelion). Earth is closest to the Sun around 3 January (147,098,074 km) and farthest away around 4 July (152,097,701 km). This results in 6 per cent more solar energy being received in January than in July. When the orbit moves to its most elliptical (aphelion), it receives 20 to 30 per cent less energy than at perihelion.

In about 10,000 to 11,000 years' time the present situation will have switched, with the Earth closer to the Sun in July and farthest away during January.

Axial precession (wobble)

As Earth moves through space, it also rotates (spins) around its axis (Figure 3.14). This action gives us day and night. This axis is not in a stable position but, like a spinning top or gyroscope slowing down, it wobbles. This is caused by the Earth not being a perfect sphere but having a bulge at the equator, and by the gravitational effects of the Sun and the Moon on the planet. The wobble takes place over a period of approximately 19,000 to 23,000 years.

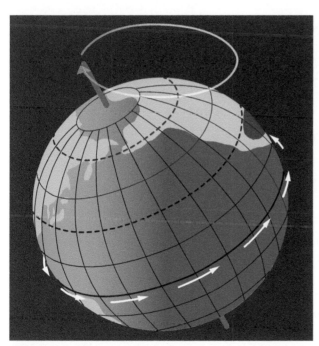

▲ **Figure 3.14** Earth's precession

Due to precession, a number of changes occur. Although at present the North Pole points almost towards Polaris (the 'north star'), the wobble will slowly move the direction in which the North Pole points. As such, Polaris will no longer be the 'north star' and other stars will take its place. In 13,000 years the pole will point towards the star Vega (Figure 3.15). Our 'north star' Polaris will not regain its position until the cycle has completed in around another 23,000 years from now.

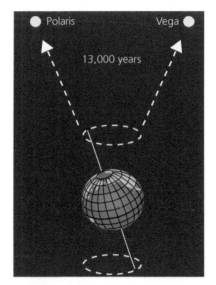

▲ **Figure 3.15** Earth's axial movement and pole position

More importantly for climate change, axial precession controls the timing of the seasons. Over time the wobble slowly changes the position of the Earth's orbit when the planet is closest to (perihelion) or farthest from (aphelion) the Sun. In truth it alters all the positions of the planet within its orbit that link to what we think of as climate.

Using the perihelion and the aphelion as our examples, we can see the large-scale changes in climate that can be caused by a precession. At the present time Earth is closest to the Sun around the time of the **winter solstice** (northern hemisphere) but in 13,000 years when the axis points towards Vega this will have swapped over and the perihelion will be during the **summer solstice**. The northern hemisphere will therefore experience winter when the planet is farthest away from the Sun and receiving less heat energy, and summer when it is closest to the Sun. This will result in greater seasonal contrasts, with the northern

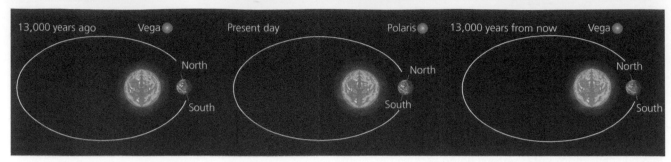

▲ **Figure 3.16** Seasonal changes due to axial precession over time

hemisphere experiencing extreme cold in winter and intense heat in summer.

It should be remembered that these changes in the position of the Earth take place over around 13,000 years and what is described above shows the extremes. During this time summer and winter will occur at different points in the orbit as part of a continuing process of climate alteration.

Axial obliquity (axial tilt)

The Earth's axis is not perpendicular but lies at an oblique angle. This tilt is what gives the planet its seasons. The northern half of the Earth (the northern hemisphere) has its summer when the North Pole is tilted towards the Sun. At the same time, the South Pole is tilted away from the Sun, bringing winter to the southern half of the planet (the southern hemisphere). Six months later the Earth has reversed this positioning, giving the north its winter and the south its summer.

The angle of tilt is not static, moving 22.3° to 24.5° every 41,000 years. At present Earth has an axial tilt of around 23.4°. The angle of tilt appears to be decreasing at present at a rate of around 119.38 cm per century.

An alteration in the angle that the planet presents itself towards the Sun would have an effect on the seasons. For example, if tilt of the axis brings more of the northern hemisphere towards the Sun, the most intense area of incoming solar radiation (the thermal equator) would move northward. This would shift the patterns of global climate northwards and create different interactions which could distort the climates further. Additional complexities would occur as the polar ice caps around the Arctic would experience increased periods and amounts of heat, causing accelerated melting, increased atmospheric moisture

and cold waters being released into the oceans. In the southern hemisphere, such a change in axial tilt could create colder conditions, extended winters and increased polar ice.

All three of the Milankovitch cycles are occurring at the same time and influencing each other. The seasons that are created by axial obliquity (tilt) are also affected by eccentricity and precession. The amount of eccentricity can accentuate the effect of precession on the heating or cooling of the hemispheres. It is suggested that variations in the Earth's orbit and movement account for only around 25 per cent of climate change.

Meteorite impact

▲ **Figure 3.17** Meteorite impact

The impact of a large **meteor** or an **asteroid** (Figure 3.17) on the surface of the Earth can result in large-scale consequences for numerous systems within the planet. By looking at the Moon, we can view many craters caused by impacts on its surface and realise that the Earth is more than likely to have been struck by objects in a similar way (Figure 3.18 on page 99).

▲ **Figure 3.18** Moon

It is believed that the Earth may even have been hit by more objects than the Moon but that erosion has worn them away. Around 170 craters have been identified on Earth (Figure 3.19), the largest being roughly 300 km in diameter. Though not undergoing the periods of extensive bombardment of the distant past, the Earth is believed to experience at least one major event, able to create a 20 km deep crater, every million years.

▲ **Figure 3.19** Meteor/Barringer Crater, Arizona USA

Extreme impact events have the ability to alter climate conditions both locally and globally. The impact explosion can have a great effect as gas, dust, aerosols and other materials are ejected into the atmosphere. Different aspects of the climate can be affected by this. The additional particles can block, reflect or scatter sunlight (albedo) and other forms of insolation causing global dimming and a resultant cooling of atmospheric and oceanic temperatures. Some aerosols at lower levels will promote the formation of clouds, resulting in additional reflection

of sunlight along with precipitation patterns and amounts being altered.

It has been suggested that the atmosphere can be corrupted by substances injected by these impacts, creating strong and damaging acid rain or a lowering of the amount of breathable air.

The alteration of topography (shape of the land) by an impact can influence climate by deflecting airstreams, changing precipitation locations or even creating blocks to, or new routeways for, oceanic currents and their warming or cooling effects.

Around 65 million years ago a meteorite of around 10 to 15 km wide, travelling at 80,400 km/h, hit the Earth with an estimated force of 100 million megatons (over 1 billion times more explosive than the atomic bombs dropped on Hiroshima and Nagasaki). The meteorite impacted near where the town of Chicxulub on the Yucatán Peninsula in Mexico is now located. The Chicxulub crater is more than 180 km wide and 20 km deep (Figure 3.20).

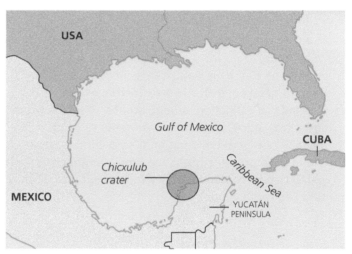

▲ **Figure 3.20** Location of the Chicxulub crater

The climate was massively disrupted by:

● the long-term effect of the vast amounts of gas, aerosols and dust in the atmosphere, blocking sunlight (10–20 per cent) and solar radiation from reaching the surface of the Earth or the lower layers of the atmosphere. This created a global 'impact winter' where average temperatures dropped by 5°C, with freezing temperatures experienced and lasting from 3 to 10 years

- the burning of forests and the deposition of ejected material all around the globe which buried plant life, allied to the reduction in the amount of sunlight and the lowering global temperatures which reduced the amount of **photosynthesis**. This resulted in the maintenance of high levels of carbon dioxide and much lower oxygen levels
- acidic aerosols blasted into the atmosphere assisting in the reduction of sunlight reaching the Earth's surface and also contributing to wide-scale acidic rainfall (sulphuric and nitric) that affected water supplies, destroyed vegetation and killed a variety of life forms
- large-scale changes to landforms (including gigantic landslips), resulting from the massive shock waves, altering existing climatic pathways and affecting atmospheric movement and oceanic currents
- the increase of carbon dioxide in the atmosphere (from the destruction of carbonate rocks and the burning of trees and other plants) which caused increased stimulation of the greenhouse effect and produced a rapid and sudden increase in global temperature once the dust had settled and the impact winter had subsided.

The Chicxulub impact happened at around the time when 60 to 70 per cent of *all* the species (plant and animal) on Earth died out. This is known as the K–T Event (Cretaceous–Tertiary **extinction event**) and marked the extinction, among others, of the dinosaurs and many plant types. It is reasonable to suggest that the violent global environmental and climatic change would have drastically changed the conditions in which these creatures survived to the extent that they suffered rapid extinction.

Volcanic activity

Volcanic eruptions can affect the climate. Materials introduced into the atmosphere alter its gas and reflective properties and affect its energy balance. In addition to this, large amounts of lava flow can alter the albedo of the surface with, for example, dark lava replacing lighter, more reflective, ice, soils, grass and other vegetation.

At present, the volcanic activity level experienced does not appear to have major long-term effects on climate but can create noticeable short-term changes. This does not rule out volcanic activity as a possible major driver in the past during periods when activity levels were higher, or rule out the possibility of future events being so. What is certain is that volcanoes can contribute to both warming and cooling of the Earth's atmosphere as well as **ozone** depletion.

Explosive volcanic eruptions eject large quantities of material into the atmosphere (gas, aerosols and ash). The most abundant gas is usually water vapour and it, along with most of the ash in the lower levels of the atmosphere, will fall rapidly to the Earth's surface within days or weeks; as such it has little effect on climate change. Materials which reach the stratosphere may remain there for several years and are quickly blown around the planet.

The second most common gas ejected is carbon dioxide, a greenhouse gas capable of trapping heat within the atmosphere. Recent volcanic activity has suggested that at present the amount of carbon dioxide released by volcanoes does not play a significant part in climate change and could not be a major driver of recent global warming. The **Intergovernmental Panel on Climate Change (IPCC)** states that since the mid-eighteenth century volcanic carbon dioxide emissions have only been around 1 per cent of the amount created by the burning of fossil fuels by humans. However, although it is not a major contributor at present, it is a contributor and different rates and extents of volcanic activity could alter this.

The emission of sulphur dioxide is being recognised by climatologists as causing the most notable climate impact resulting from volcanic eruptions. When large-scale eruptions push this gas into the stratosphere, it combines with oxygen and water to form sulphuric acid and condenses to form sulphate aerosol droplets. Vast amounts of these minuscule aerosol particles are formed following volcanic eruptions and they grow, forming a dense layer of haze which may remain in the stratosphere for several years. The aerosols intercept incoming solar radiation and heat themselves and the surrounding atmosphere. This interception stops the energy from reaching lower levels of the atmosphere and the Earth's surface. At the same time, these aerosols can also directly reflect incoming radiation back out into space.

Due to this absorption and reflection the complex system of climate creation is adjusted and this leads to climate change. The aerosols directly create a cooling of the lower atmosphere and planet surface as their presence has a larger effect than that of the added carbon dioxide and any additional warming it could provide. As such, the planet experiences a short period of cooling. Recent studies have shown this cooling to be around 0.2°C, resulting from a reduction of about 5–10 per cent of energy received at the Earth's surface.

The cooling effect of sulphur dioxide from one large-scale eruption lasts for two to four years. This relatively short consequence is due to the aerosols eventually falling from the stratosphere to be dispersed by winds and rains lower in the atmosphere.

Studies of historical and more recent eruptions seem to back up the belief it is sulphur dioxide and not ash that is responsible for this climatic cooling. The Indonesian eruptions at Tambora (1815), Krakatau (1883) and Agung (1963) all resulted in surface temperature reductions of around 0.18–1.3°C. When the amount of ash and sulphur injected into the atmosphere for each eruption was compared, the differences observed suggested that it was the amount of sulphur dioxide being added that had the most effect. This and following investigations have shown that it is not how explosive a volcano is but the amount of sulphur dioxide being released that affects climate change the most.

Depletion of ozone

An additional impact of the sulphate aerosols is to assist in the depletion of ozone in the stratosphere. The surface of these aerosols assists in the chemical reactions to alter nitrogen and chlorine in the stratosphere and these, in turn, destroy ozone. Reduction of the ozone layer allows more ultraviolet light to reach the lower layers of the atmosphere and the surface of the planet. The additional energy of ultraviolet light can alter the climatic system and its responses. It also endangers life near the surface. Damage to vegetation reduces oxygen creation and allows for an increase in carbon dioxide in the atmosphere. The removal of vegetation alters the reflective properties of the surface and the temperature of the air above it. After the eruption of Mount Pinatubo in the Philippines in 1991, ozone levels in the mid-latitudes were reduced by 5–8 per cent.

Acid rain

At lower levels in the atmosphere, the sulphur dioxide mixes with the oxygen and water in the atmosphere to create acid rain. This in itself is an alteration to the climatic quality but it can have further impacts. The addition of acid rain to vegetation can cause its destruction or at least a reduction in its ability to thrive, resulting in changes to the oxygen/carbon dioxide balance and even surface albedo. The main ecological events can be seen in water environments where the water increases in acidity. This would affect rivers, lakes, seas and oceans. There would follow a cascade of events, reducing the number of biological organisms and oxygen-generating capabilities.

Gas release at the Earth's crust

Although most of what has been discussed here is related to explosive volcanism at the surface, there are many different ways that gases may be released. In volcanic areas gases can be released through cracks in the Earth's crust such as fumaroles. These may emit steam, carbon dioxide, hydrogen sulphide and hydrogen chloride into the atmosphere. Volcanic action on the ocean bed also releases gases into the oceans and eventually the atmosphere, which can change their gas contents and properties and so affect the climate. The exact amount of influence these submarine emissions have on climate has yet to be clarified.

During the recent research along the Mariana Arc in the North Pacific Ocean, a hydrothermal vent was discovered emitting gas bubbles. In 2006, samples of the bubbles from a 1.6 cm² section of the vent were analysed, identifying carbon dioxide as the main gas contained. From the amount of gas captured it was worked out that if this rate of emission remained constant, this very small section would release around 9 million tonnes of carbon dioxide every year. Although every vent does not release the same amount of carbon dioxide, it is interesting to think that if this was the average for every 1.6 cm² section of every vent around the world's 22,000 km of volcanic arcs, the amount of carbon dioxide added to the oceans, and eventually the atmosphere, would be extremely significant.

Impact of flood basalt eruptions: Siberian Traps

Earlier it was mentioned that recent volcanic activity does not appear to have a massive effect on present-day climate or long-term conditions. There is, however, evidence that in the past there have been events of such magnitude and longevity that they have had a massive effect on the climate. Although a variety of volcanic eruption types were found during these periods, it is believed that flood basalt eruptions, which spray and pour out large amounts of lava continually over an extremely long period of time, were responsible (Figure 3.21). The most common rock type erupted is basalt and this tends to run or pour out over prolonged periods, with gases being exuded into the atmosphere.

▲ **Figure 3.21** Lava being ejected from a fissure and flowing over a large area of land

In northern Russia there is an area known as the Siberian Traps (Figure 3.22). The word 'traps' comes from a Swedish word meaning steps or stairs and this describes the step-like hills created by flood basalts. This Siberian landscape is what remains of a massive area (thousands of square kilometres) of flood basalt eruptions that occurred in northern Pangaea some 251 to 252 million years ago (Figure 3.23). These eruptions are believed to have continued for *at least* 2 million years and covered an area about the size of Europe in lava. It is estimated that a total volume of 3 million km³ of basalt poured out.

Palaeoclimatologists have suggested that the Siberian Traps eruptions were responsible for the massive climate and environmental changes that triggered Earth's greatest mass extinction event (the P–Tr Event, or the Permian–Triassic extinction event) which also took place around 251–252 million years ago. Around 96 per cent of marine and 70 per cent of terrestrial vertebrates became extinct during this period.

It is thought that this colossal and long-lasting volcanic episode triggered events that massively changed the Earth's climate and so created the conditions for the mass extinctions. Flood basalt eruptions discharge large amounts of gases such as carbon dioxide, sulphur dioxide and hydrogen sulphide. At first it was thought that these would have acted as aerosols and that the planet would have become substantially colder for an extended period, but proxies suggest that there was, in fact, an overall heating of the atmosphere.

It is generally accepted that atmospheric temperatures rose by around 10°C from the previous level and that this would have been enough to trigger

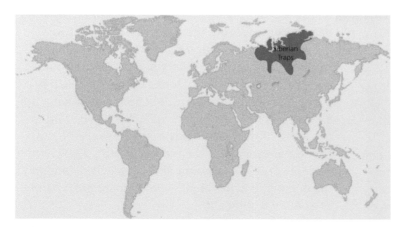

▲ **Figure 3.22** Location of the Siberian Traps today

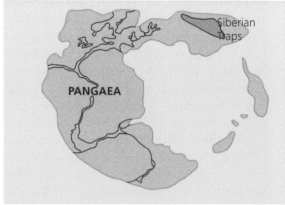

▲ **Figure 3.23** Location of the Siberian Traps in Pangaea

the mass extinctions. Recent research explains this situation by suggesting that in the early stages of the event the aerosols entered the lower stratosphere and reduced the amount of light/energy reaching the Earth's surface (global dimming). This reduced the temperature and created a volcanic winter, polar ice caps formed and sea levels were dramatically reduced. Some suggested that the magma plume pushing up from below had the effect of raising the crust high enough to allow full glacial ice (possibly even ice sheets) to form in areas that surrounded the volcanic activity. This would also have resulted in increasing aridity, a variance in ocean circulation and an increased albedo as the more reflective white snow and ice covered the previously darker landscapes.

The reduced sunlight and temperature severely decreased the amount of plant life and its ability to remove carbon dioxide from the atmosphere and to create oxygen. Over the millennia the amounts of carbon dioxide built up in the atmosphere and oceans, resulting in an extreme greenhouse effect encouraging temperature increase. The carbon dioxide would have blanketed the planet and have been at such a level that it overpowered any cooling effects.

Some palaeoclimatologists are not convinced that this was the only cause. Calculations based on the gases that would have been given off by the volume of lava produced were fed into computerised climate models. The results state that the temperature increase would only have been around 5°C. This temperature increase is not believed to have been enough to kill off so many of the species. An additional element triggered by the volcanic activity and the rising temperatures is now believed to have played an important part.

Impact of underwater methane gas

The release of underwater methane gas is now being recognised as an additional driver of climate change through its effect on ocean and atmospheric conditions. This gas may come from underwater volcanic action or from methane held in crystal structures (clathrates) on the ocean floor. Sudden and large-scale/long-term release of methane (a greenhouse gas) from either of these is suggested as the cause of increased ocean acidification (OA), increased ocean temperature and heightened atmospheric temperatures. Scientists believe that

the increase in sea temperatures during this period activated the release of methane gas from clathrates and this is calculated to have been able to stimulate a runaway greenhouse effect and increase atmospheric temperatures by up to 6°C. It has been calculated that an increase of this amount, if it were to happen today, would raise Scotland's temperatures to the equivalent of those in the Sahara Desert in northern Africa. The additional temperature rise from the increase in methane would bring the total increase to around that identified from proxies.

All of these changes would have been accompanied by the previously established effects of a volcanic eruption (such as acid rain and ozone depletion), which would have exacerbated the change in climate and the chances of mass extinction.

Impact of flood basalt eruptions: Deccan Traps

At the time of the K-T extinction event of 65 million years ago, previously proposed as being caused by a meteorite impact, there was another large-scale flood basalt eruption taking place at the Deccan Traps in India (Figure 3.24). Most of the basalt was erupted between 65 and 60 million years ago when India was an island continent slowly moving northwards to its present position.

Like the Siberian Traps, this massive and long-term volcanic event caused wide-scale disruption to global climate systems.

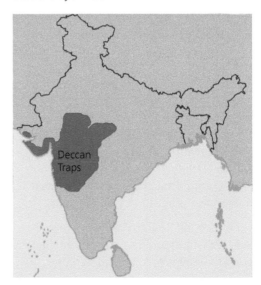

▲ **Figure 3.24** Location of the Deccan Traps today

Research opportunity

We've already looked at the Siberian Traps so this is an opportunity for you to look into the Deccan Traps.
To start with, try an internet search using the following keywords and phrases:

- Deccan Traps   • Deccan Traps and climate change   • Deccan Traps and atmospheric gas balance
- Deccan Traps and mass extinction   • Deccan Traps and environmental change.

Tambora and Krakatau eruptions

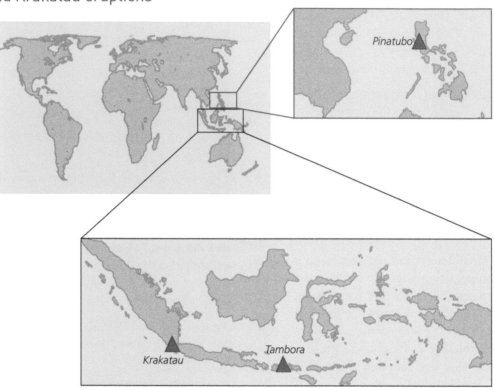

▲ **Figure 3.25** Locations of Mounts Tambora, Krakatau and Pinatubo

The Indonesian eruptions at Tambora (1815) and Krakatau (1883) happened at a time when scientific investigation was becoming better organised. Early weather instruments and records allow for comparisons before, during and after these events. Tambora erupted several times in the period from 1812 to 1815 but the explosive eruption starting on 10 April 1815 was exceptional. This was the largest volcanic eruption on the planet for 10,000 years.

It is estimated that around 160 km³ of lava blasted out from the volcano. A giant eruption column reached an estimated height of 40–50 km, injecting ash particles and aerosols into the stratosphere. For two days, and up to a distance of approximately 600 km, the erupted material completely blocked out sunlight. Barometers recorded a pressure wave that passed around the world several times.

Although much of the ash fell from the atmosphere within a fortnight, the tinier aerosols remained in the atmosphere for a number of years. It is estimated that Tambora released around 50 to 120 million tonnes of sulphur dioxide into the stratosphere and that the majority of aerosols formed from this as droplets of **sulphurous acid**. Winds transported the aerosols around the globe, creating a veil that was to block or reflect both heat and light from the Sun and reduce global temperatures by around 0.5°C over a one- to two-year period.

It is estimated that locally the direct results of the eruption from lava, multiple tsunamis and destruction of farmland leading to famine killed around 60,000 people. The quality of air to breathe was reduced due to high sulphur and other toxic gas content. Inhalation of these gases led to people becoming sick and an increasing number of lung diseases.

The global effect was even more severe. Such was the amount of cooling caused by Tambora that 1816 became known as 'the year without a summer', vastly reducing crop yields, killing animals and causing a major **starvation** event in the northern hemisphere. An estimated 100,000 people died as the result of famine and disease. Among the unusual weather phenomena experienced during this short 'volcanic winter' were June frosts and heavy snow in the USA and Canada, extreme rainstorms causing flooding in Europe, brown- and red-coloured snow in Hungary and Italy, and the blocking of the monsoon in both India and China, denying crops rainfall when it was needed.

Research opportunity

What is described above is a very brief version of the eruption and effects of Tambora. Try to find out more about the situation and its wide-scale effects on climate and in particular 'the year without a summer'. To start with, try an internet search using the following keywords and phrases:

- Tambora
- Tambora 1815 and climate change
- Tambora and climate phenomena
- Tambora and sunsets
- effects of the 1815 Mount Tambora eruption across the globe
- Tambora and monsoons
- Tambora Blast from the Past
- Tambora and San Francisco fog
- Tambora 1815 and J.M.W. Turner
- Tambora 1815 and Frankenstein.

To follow up on Tambora it would be interesting to find out about the 1883 Krakatau (Krakatoa) eruption, also in Indonesia. Although less powerful than Tambora, Krakatau also led to changes in global climates, alterations to precipitation patterns, increased acid rain and optical phenomena such as a lavender-coloured Sun and moonbeams glowing red or green.

More recent volcanic events

▲ **Figure 3.26** Mount Pinatubo erupts, 1991

In 1991 Mount Pinatubo in the Philippines (Figures 3.25 and 3.26) erupted somewhere between 13 million and 27 million tonnes of sulphur dioxide into the stratosphere. Over 90 per cent of this was emitted during the nine hours of the major eruption on 15 June. Within two hours of the eruption starting, the ash and gas plume had risen through the tropopause (17 km above the Philippines at that time) to a height of around 35 km and had spread over 400 km wide. Elements of this cloud were to last for a number of years within the upper atmosphere.

Because the eruption occurred in the modern era, scientists were able to monitor the effects of the eruption with instruments at ground level, on aircraft and from satellites. Weather satellites tracked the ash cloud for four days until enough ash had fallen back to Earth to make it difficult to identify. After this a satellite equipped with a **Total Ozone Mapping Spectrometer** (**TOMS**), with the ability to measure and follow sulphur

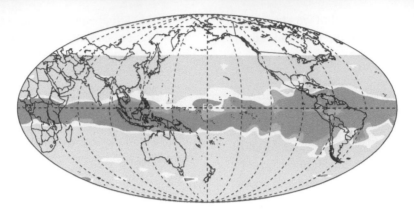

▲ **Figure 3.27** Map of aerosol plume from Mount Pinatubo, 1991

dioxide emissions from volcanoes, monitored the path of the sulphur dioxide cloud. In two weeks the aerosol cloud spread completely around the Earth and within a year had expanded to cover the whole planet (Figure 3.27). It is estimated that the sulphur dioxide that reached the stratosphere converted into sulphuric acid aerosols within a short period of around 28 days, during which time the cloud had travelled around the planet.

Scientists observed as the volcanic haze of sulphate aerosols reflected and absorbed incoming solar radiation. Measurements taken in 1992 suggest that there was a reduction of 25–30 per cent in the level of incoming solar radiation reaching the lower atmosphere and the planet's surface. The dense aerosol cloud remained in concentrations much higher than before the eruption for two to three years. As a result, during 1991–93 global temperatures were temporarily reduced by around 0.5°C (Figure 3.28).

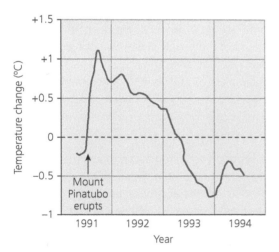

▲ **Figure 3.28** Graph of global temperature drop after the Mount Pinatubo eruption

The effect of emitted sulphate aerosols on the ozone layer was also identified. Using records collected since 1956 and by satellites since the early 1970s for comparison:

- The average global ozone amount reduced to 2–3 per cent in 1992–93, lower than any previous year.
- The planet's mid-latitudes ozone levels dropped to their lowest levels in 1992–93.
- In 1992 in Antarctica the fastest ever rates of ozone depletion were recorded.
- In 1992 the southern hemisphere **ozone hole** expanded to its greatest extent *to date* of around 24.9 million km^2.

Additional effects attributed to the Pinatubo eruption included:

- 1992 – third coldest and third wettest summer in the USA for 77 years
- 1993 – extreme drought in the Sahel area of Africa
- 1993 – flooding along the Mississippi River in the USA.

The above is a very simplistic representation of the effects of Pinatubo on weather and climate, highlighting what is accepted thinking. With the advancement of scientific analysis, a more complex picture of the effects of Pinatubo has arisen and highlighted the intricacies of the planet's climate-generating system. Although overall global temperature decreased, the cooling was not spatially uniform and some areas experienced higher temperatures than the norm. The lower equatorial stratosphere experienced significant warming which, in turn, increased the equator-to-pole temperature gradient in the southern hemisphere. This influenced climate pattern and weather intensity. A knock-on effect of this, in the northern hemisphere, was to produce an alteration of atmospheric circulation patterns that manifested itself in increased surface warming over northern and eastern Europe and Siberia. There was also a cooling over Greenland, the Middle East and the Mediterranean.

Research opportunity

A volcanic eruption can result in the blast destroying the top of a volcano. Conduct some research to find out how extreme reduction of height could influence local climatic patterns.

Plate tectonic movement

The lithosphere (the generally solid outer layer of Earth) is divided into sections known as **tectonic plates**. These plates are divided from each other by large faults (cracks) in the Earth's crust and fit together like a jigsaw puzzle. These faults provide areas of weakness which allow for movement of the plates and release materials from below the crust onto the surface. The plates slide over the partially molten layer below them. Due to this the Earth's crust is not static but moves, reshapes, is depleted and replenished. These processes are powered by the internal thermal movements of **magma** below the crust. The movement of these plates is very slow, with the global *approximate average* of movement being about 3 cm per year at present.

It is important to realise that plate tectonic movement continually happens, much of it at a very slow pace, so that we are not aware of its effects – for example, the continuing growth of the Himalayan mountains (see page 109). At other times volcanoes triggered by the movements may have a relatively quick effect on weather patterns and climate.

Research opportunity

This is not the place for a great amount of detail about plate tectonics so if you are concerned that you don't know enough about the subject, spend some time finding out more. To start with, try an internet search using the following keywords and phrases:

- plate tectonics
- evidence for plate tectonics
- plate tectonics and glaciation.

Plate tectonic movement and volcanicity

That plate tectonics drives climate change may be seen by the volcanic processes associated with it. Volcanic processes are an integral part of plate tectonic movement. Put very simply, volcanoes mostly occur in belts running along the major fault lines between plates. Where plates are moving apart (at a **constructive boundary**), magma, along with large amounts of volcanic gases, rushes to the surface to fill the gap, creating volcanoes and new crust. Where less dense crust is forced downwards below another plate that it has collided with (at a **destructive boundary**), it melts and the molten rock and gases rush to the surface to form volcanoes. Volcanic plumes can also be forced to the surface, breaching the crust to form new land (as is being experienced in Hawaii; we have already read about the Siberian Traps and the Deccan Traps).

All of the climate-influencing changes discussed in the section on volcanoes above should be seen as resulting from plate tectonics. During periods of large-scale plate tectonic movement, increased volcanic activity tends to release vast amounts of carbon dioxide into the atmosphere resulting in global temperatures rising due to stimulation of the greenhouse effect.

Plate tectonic movement and location of landmasses

Most of us are familiar with how Earth looks on a map; we can recognise the position of continents, oceans and even mountain ranges. But if we were to look at a map showing Earth 250 million years ago it would look very different. There would be only one continent, a super-continent which we now call Pangaea, and one super-ocean known as Panthalassa. Since then Pangaea has split apart and the continents created have slowly drifted to their present positions. The movements disrupt ocean currents, create mountains and generally change the climatic conditions of the planet.

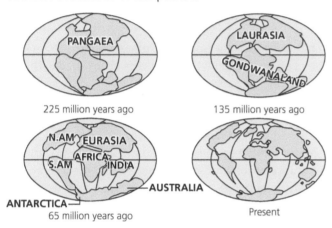

225 million years ago   135 million years ago

65 million years ago   Present

▲ **Figure 3.29** Continental movement through time

As plates move the ocean beds, the continental landmasses of which they are comprised shift position on the globe. If a continental plate moves to higher latitudes, it will receive less of the incoming solar radiation and its climate will cool. If it moves to lower latitudes, temperatures will rise. These movements also change the seasonal qualities and by crossing from one side of the equator to the other, this would prompt a reversal of seasons. Both oceanic and landmass temperatures would be affected, changing the qualities of both to further stimulate climate change and weather patterns. As tectonic plate movements are very slow, these changes only become significantly observable over geologic time.

The amount of energy reflected by different surfaces (albedo) also affects the global climate. The position of landmasses and oceans can affect the amount of energy that is reflected back out to space and is lost to the climate system. A landmass with, for example, a tropical desert will reflect around 25 per cent of incoming solar radiation whereas a tropical ocean

located in the same position would reflect only 7 per cent. Although a landmass would still have high temperatures and tropical or equatorial regions, much of the energy received would eventually be released back into space, whereas an ocean in the same position would maintain the energy and transfer this around the globe.

Part of the climate system relies on the evaporation of water to transfer heat from surface level into the atmosphere. When in the atmosphere, this heat energy acts as an energy source to assist in the powering of weather systems around the planet. With landmasses occupying the lower latitudes, near the equator, less water is available for evaporation and as such less energy is available to power the weather systems.

At high latitudes incoming solar radiation reduces. This encourages much lower temperatures and snow and ice accumulate on landmasses. This creates a **positive feedback effect** whereby the intensified albedo (by 65–80 per cent) due to the white surface further reduces temperatures and creates circumstances that allow increased accumulation and spreading of snow and ice. The spreading again increases the area of higher albedo and encourages continued expansion of the snow and ice. So, due to an increase in continental landmass at high latitudes, the area and thickness of permanent ice cover extends and the planetary albedo is increased; this in turn results in global cooling.

Even waves in the atmosphere such as the **jet stream** can be altered due to the conditions beneath them. With land and oceans heating differently, the atmospheric conditions above them are also altered, assisting in the movement/location of the jet stream.

Plate tectonic movement and increased altitude of land

Simply by land occupying an area that was previously ocean there is some change to wind patterns. This may be the wind altering in altitude as it meets landforms or, to a greater extreme, being blocked or diverted by mountain ranges. The location of mountain ranges, their length and height, can interfere with even the major wind belts resulting from atmospheric circulation. Along with physical blocking, the pressure differences across mountain ranges can strongly influence the strength, direction and distribution of the winds.

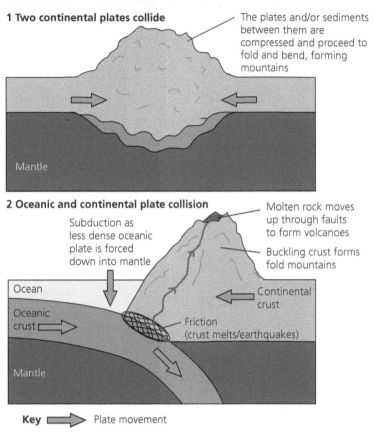

1 Two continental plates collide

The plates and/or sediments between them are compressed and proceed to fold and bend, forming mountains

Mantle

2 Oceanic and continental plate collision

Subduction as less dense oceanic plate is forced down into mantle

Molten rock moves up through faults to form volcanoes

Buckling crust forms fold mountains

Ocean

Oceanic crust

Continental crust

Friction (crust melts/earthquakes)

Mantle

Key ➡ Plate movement

▲ **Figure 3.30** Plate movement and mountain building

Warm, moist winds meeting mountains are forced to rise, resulting in the air cooling and moisture condensing and falling as precipitation. Dropping back down over the mountain range the air is dry and as it warms it sucks up the moisture from the surrounding area. This results in what is called a rain shadow area on the leeward side of mountains (the other side from where the wind is coming) which has dry conditions. It should also be recognised that the precipitation falling on the mountains may also be in the form of snow due to the lower temperatures at altitude and this could encourage additional snowfield or glacial expansion.

Plate tectonic movement can also result in high land/ mountain ranges being formed (Figure 3.30). When plates converge and push against each other, the crust becomes bent and folded (crumpled up) and this squeezes land upwards. The consequence of this is mountain building (**orogenesis**) and the resulting mountains are known as fold mountains.

Around 40 million years ago the Indian tectonic plate, which had been moving steadily northwards at around 15 cm a year, crashed into the Asian plate and this collision created a crumple zone of folded crust. As the movement continued, the upward folds became increasingly high and gradually formed the Himalayan mountain range (Figure 3.31): 2400 km in length and covering 1,089,133 km². With an average height of over 5000 m, there are more than 100 mountains above 7200 m, including the

▲ **Figure 3.31** Himalayan mountains

highest land-based mountain on the planet, Mount Everest, at 8850 m above sea level. Recent studies have shown that the Himalayan mountains are still growing higher at a rate of 6.1 cm per year due to the continuing collision between the plates.

Large rock surface areas (including some that were previously on the ocean floor) were exposed to the atmosphere, causing an increase in chemical weathering which removed carbon dioxide from the atmosphere. The vast amount of rock raised up had a large enough effect on the atmosphere's carbon dioxide levels that global temperatures were cooled. The land, uplifted into cooler altitudes, also encouraged glacier formation and snow build-up. As described earlier, this in turn affected the albedo of the landscape and created a feedback event, encouraging continued expansion of the glacial area.

The creation of the Himalayas has had a massive effect on the climate. The Himalayas act as a barrier to the systems of air and water circulation and help to create the climatic and meteorological conditions to their north and south. During the winter months, cold continental air coming across the plains from Siberia and central Asia is blocked from reaching India and south Asia to the south. This keeps India and south Asia warmer than other regions at similar latitude. During the monsoon season, warm, moist, south-westerly winds moving from the Indian Ocean upwards across India are blocked by the Himalayas and forced to rise. As they rise, the air cools and this causes the moisture to condense and fall as heavy precipitation (rain and snow) on the southern side of the mountains. As the air passes over the mountains, most of the moisture has been removed and to the north conditions are arid. The average precipitation on the southern slopes of the Himalayas is between 1530 mm and 3050 mm. To the north, precipitation levels may be as low as 75–150 mm. This leaves the area to the south of the mountain range with higher temperatures and an increased amount of seasonal rainfall, while to the north the land is much drier and experiences exceptionally cold winters.

Plate tectonic movement and oceanic circulation change

Oceanic circulation plays a major part in the Earth's climate. Due to the Earth being a sphere, incoming solar radiation does not heat the Earth evenly. The curvature of the Earth means that equatorial regions receive more intense heating than polar regions. Along with the atmosphere, the oceans are part of a system which tries to equalise some of the inequalities by redistributing heat towards the poles and returning cooler conditions in the opposite direction.

Research opportunity

We have provided a fairly simplistic explanation of the action of the atmosphere and oceans to redistribute heat. If you feel you don't know enough about this subject, this is your chance to find out more! If you have already completed the Physical Environments: Atmosphere section, you may wish to refresh your memory from your notes. Also try an internet search using the following keywords and phrases:
- global heat budget • global heat transfer
- oceanic circulation.

Plate tectonic movement and its associated volcanicity has the ability not only to alter the pattern of oceanic circulation but to deny access to former oceans. With its movement northwards, the continental plate on which Africa sits entrapped what would become the Mediterranean Sea. But at some point between 6 and 5.3 million years ago connections between this sea and the oceans were cut off. It is believed that this was related to an increased rate of ocean spreading (new oceanic floor being created at a constructive plate margin) far away in the centre of the Atlantic that forced the crust to be squeezed upwards around what is now the Straits of Gibraltar. This created a block, stopping oceanic water from entering the Mediterranean. Due to high levels of water evaporation (around three times more than the input from any rivers or precipitation), in as little as 2000 years the sea dried out and changed into a hot and salty desert with the surrounding areas also denied the cooling effects and moisture from the sea.

An example of an oceanic current is the North Atlantic Drift, where warm waters move north-eastwards from the Caribbean and moderate the effects of latitude on, for example, Scotland. The effect of the warming from this current can be seen when comparing Glasgow (Scotland) with Moscow (Russia). Both of these cities are very close to being at the same latitude: Glasgow 55°51'N and Moscow 55°45'N. Being in a location far from the oceans, Moscow does not experience the warming influence of an oceanic current in the way that Glasgow does. The mitigating effect of the North Atlantic Drift is particularly evident through analysis of winter climate statistics. Glasgow's average temperatures are kept above freezing while Moscow's drop to as low as −9°C.

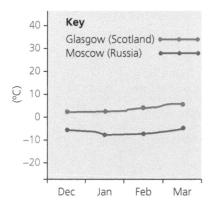

▲ **Figure 3.32** Comparative average winter temperatures, Glasgow and Moscow

If this warming oceanic current were to be 'switched off', it is not unreasonable to suggest that Scotland could see winter temperatures similar to those of Moscow. Removal of the North Atlantic Drift would also result in the formation of sea ice, similar to the most extreme winter Baltic Sea conditions, and even icebergs off our shores! The history of changes to

oceanic patterns has shown that they do alter climatic conditions but, once again, the complexity of climate change is demonstrated in that even the deflection of warmer currents into areas does not always help to increase temperatures overall.

A massive global climate change was triggered around 3 to 5 million years ago as North and South America moved towards each other. This movement created the Isthmus of Panama which joined the two continents together. The isthmus blocked what had previously been the pathway of a warm current that flowed around the equator in what had been a single ocean. With this route blocked, the warm waters were forced to change direction and travel polewards in the northern hemisphere. This was the creation of what we referred to above as the North Atlantic Drift.

Instead of increasing temperatures at higher latitudes, what happened was a disturbance of the climate that resulted in the creation of an ice age. What the current did was to bring additional moisture to the northern regions which evaporated into the atmosphere. Due to the cold conditions, this supplementary atmospheric water was precipitated as snow. Extended snow coverage increased the albedo of the land area and so encouraged a further increase in temperature loss. Due to this, the climate became colder, there was more snow and ice and the process became self-stimulating. This triggered global cooling.

 Task

1 Create a mind map (or summary spider diagram) showing the major physical (natural) drivers of climate change and how they influence climate. Remember to make the mind map fit how you learn. Do you like to use colours? How many different branches should your map have? Don't just stick to one page or one shape – do what you need to do!
2 What do the following terms mean and why are they important to the Earth's climate?
   a) Insolation
   b) Albedo
3 How do sunspots affect the climate of the planet?
4 What are aerosols and how do they affect the level of incoming solar energy?
5 Explain the impact of Mount Pinatubo on global warming in the 1990s.
6 Why is it believed that modern volcanic eruptions only have a short-term effect on the climate?
7 How does an increase in the altitude of land (as created by plate tectonic movement) affect the climate? Refer to at least one actual example.

Summary

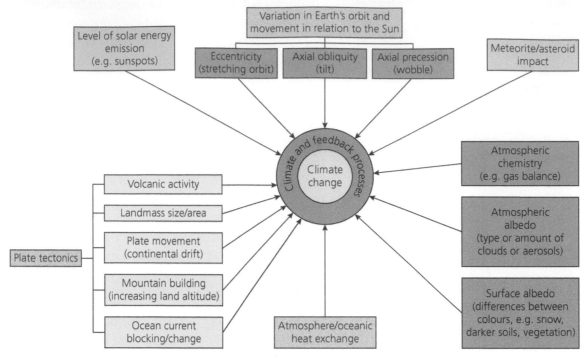

▲ **Figure 3.33** Physical drivers of climate change

The Earth's climate has changed dramatically in the past without the influence of human beings. The physical drivers that created these changes still exist and have the potential to influence or change the climate. They have shown the ability to massively alter climate on a local and global scale, destroy environments, create ice ages and eradicate species.

Individually these natural drivers can change climate, but their effects are magnified by the complexity of the climate system itself and the feedback systems contained within it.

These natural drivers have shown that they have the influence and power to stimulate any changes in climate observed at present. However, scientific analysis of data collected suggests that there is a more dominant influence at present: that of human activity.

Anthropogenic (human) drivers of the climate system (anthropogenic forcings)

Human beings, with their utilisation of resources and adaptation of the environment, interact with the planet's climate mechanisms. Human influence on climate had minimal impact until the last 200 years.

However, in this short time span, industrialisation and a rapid increase in population have amplified the ability of human actions to force change.

The main anthropogenic influence has been on the greenhouse effect and atmospheric chemistry. We have already seen how the greenhouse effect benefits the planet by trapping heat that would have normally escaped into space and thus helps to maintain the conditions of warmth that allow life to survive on the Earth. The heat is trapped by gases referred to as greenhouse gases (for example, carbon dioxide, methane, ozone and water vapour) and alteration in the amount of these in the atmosphere can lead to increases or decreases in global temperatures. Increased amounts of greenhouse gases result in additional heat being trapped and the planet warming up.

The influence of human beings has resulted in the term 'enhanced greenhouse effect' being used to describe the greenhouse effect in relation to global warming in the modern era (since the Industrial Revolution).

Human activities have added massive amounts of greenhouse gases to the atmosphere, creating an observable change in the planet's atmospheric gas balance as well as a continuing increase in average global temperatures. The level of greenhouse gases in the atmosphere remained relatively stable in the 10,000 years up to the beginning of the Industrial Revolution (1750). Levels of atmospheric carbon dioxide alone increased from 280 ppm (parts per million) to over 405 ppm in just under 270 years. The concentration of carbon dioxide in the atmosphere has now been increased to a level higher than it has been for at least 800,000 years.

Sources of human activity-related emissions

The largest sources of human activity-related emissions of greenhouse gases come from industry, electricity and heat generation, agriculture, transport and the making of cement (Figure 3.34). Industrial production and the generation of energy to sustain modern lifestyles remain widely dependent on the burning of fossil fuels (coal, natural gas, oil and petroleum). These emit a number of greenhouse gases but in particular carbon dioxide, which is believed to be responsible for around 75 per cent of the warming from **anthropogenic greenhouse gas emissions** at present.

Carbon dioxide

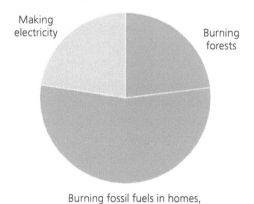

▲ **Figure 3.34** Sources of carbon dioxide contribution to greenhouse gases

Impact of industry

Although industries are responsible for the use of massive amounts of fossil fuels, the role of the individual and domestic use should not be ignored. Many products that we use have been manufactured, such as plastics that have used oil in their creation. Every time we switch on a plug, television set, computer, light, kettle, washing machine or turn up

the heating, the majority of energy that allows us to do so has come from the burning of fossil fuels. And let's not forget car, bus, train and plane journeys and all those products in our shops and supermarkets that have been transported, not just locally but from around the world.

At the start of the eighteenth century, carbon emissions were negligible, but within only 312 years the annual total had hit an all-time high of 8.7 billion tonnes being added to the atmosphere (Figure 3.35). Carbon dioxide builds up in the atmosphere because it stays there for around 200 years. In the section on physical drivers we looked at the Deccan Traps and their wide-scale disruption to global climate and ecosystems. Compared with current human output, it has been estimated that the average annual output of carbon dioxide by the Deccan Traps was a mere 60 million tonnes per year. Present-day volcanic activity releases around 135 times less carbon dioxide than human activity.

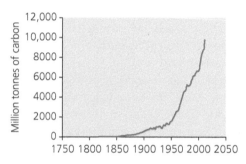

▲ **Figure 3.35** Global carbon dioxide emissions from fossil fuel burning, 1750–2018

The rate of temperature increase has risen throughout the last 200 years, with a notable dramatic rise from 1980 onwards. Comparison of the data in Figures 3.35 and 3.36 shows a close relationship between the patterns of carbon emissions into the atmosphere and the increase of global temperatures.

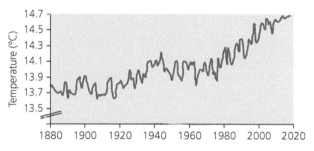

▲ **Figure 3.36** Average global temperatures, 1880–2018

Impact of deforestation

The rapid expansion of global population has stimulated use of resources (through a need for space, food supplies, construction materials and power-generating materials). There has also been massive change in land use, for example, with vast areas of forest being cleared to access resources, create farmland and living space. This produces a number of effects.

In terms of carbon dioxide, the clearance of forests on a vast scale reduces the effectiveness of one of the natural **carbon sinks**. These naturally accumulate and store carbon chemicals and one of these is by vegetation through the process of photosynthesis. In this process carbon dioxide is removed from the atmosphere because carbon is trapped within the plants and oxygen is released. At present over 30 per cent of the planet's land area (just under 40 million km²) is covered by forests. It is estimated that around 130,000 km² is destroyed every year, though it should be noted that the overall loss is being reduced due to efforts to replant or to preserve natural forests. The net levels of loss are still alarming, at around 70,000 km² per annum. In the Amazon rainforest in South America alone around 19 per cent has been lost in the last 50 years, mostly due to the creation of grazing land for cattle ranching to supply the expanding world market for beef.

Some estimates suggest that forests worldwide store around 600 gigatonnes of carbon within them. Deforestation accounts for the emission of around 10 per cent of the world's greenhouse gases. Much of this comes from the burning of trees and other vegetation during clearance.

Impact of livestock rearing

Globally, livestock rearing (including poultry) accounts for 14–17 per cent of anthropogenic greenhouse gas emissions (Table 3.2). These figures include the growing and processing of animal feed as well as levels of flatulence from the animals themselves. Increased numbers of livestock result in additional emissions.

Worldwide, livestock rearing creates approximately 6 billion tonnes of greenhouse gases per year. Of

Table 3.2 Carbon dioxide emissions resulting from meat production

Production of 1 kg of meat	Produces CO_2 amounting to
Beef	34.6 kg
Lamb	17.4 kg
Pork	6.4 kg
Chicken	4.6 kg

these, around 1.7 to 2.9 billion tonnes of methane are produced through the livestock's flatulence and excrement (faeces).

Seventy-six per cent of the UK's nitrous oxide (N_2O) emissions come from agriculture. The gas is mostly emitted from inorganic nitrogen fertilisers (used in the growing of crops including animal feed) and stored manure.

Impact of cement manufacturing

The manufacture of cement is responsible for 8 per cent of all anthropogenic emissions of carbon dioxide. Cement is a major ingredient in concrete which is the world's primary building material and is indispensable for large-scale construction. It is believed to be the second most 'consumed' substance in human activity after water. By weight, the worldwide use of concrete is more than twice that of plastics, wood, aluminium and steel combined.

Production of cement releases carbon dioxide in two ways, one directly and the other indirectly. Direct emissions come from the chemical process known as **calcination**. To create cement, limestone (calcium carbonate), which is composed of skeletal fragments of marine organisms and acts as a store for carbon, is heated to high temperatures and breaks down to form calcium oxide and carbon dioxide. Indirect emissions come from the fossil fuels burned to heat the process, production of electricity for additional machinery and through transportation of the final product (Figure 3.37 on page 115). The burning of fossil fuels to produce a tonne of cement generates the equivalent weight of carbon dioxide.

It has been estimated that annual emissions from cement production will have to be reduced by at least 16 per cent by 2030 if climate change is to be tackled.

Key

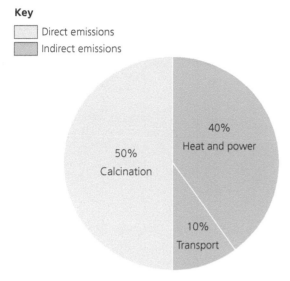

◻ Direct emissions
▨ Indirect emissions

50%
Calcination

40%
Heat and power

10%
Transport

▲ **Figure 3.37** Carbon dioxide emissions from cement manufacturing

Impact of additional carbon dioxide in the atmosphere

Although carbon dioxide is a natural part of the Earth's atmosphere and forms part of the planet's **carbon cycle**, the actions of humans adding massive amounts of the gas may not only directly, but also indirectly, affect climate. The carbon cycle depends on a system that circulates carbon among the atmosphere, oceans, vegetation, soils and animals.

The additional carbon dioxide plus the removal or alteration of elements of the cycle can push it out of balance and create positive feedback events too. By reducing forests, more carbon dioxide remains in its gas form in the atmosphere and adds to the greenhouse effect. Increasing temperatures encourage increased evaporation and **evapotranspiration** and so add more water vapour (the biggest contributor to the greenhouse effect) to the atmosphere. Increasing temperatures stimulate thermal expansion of water and this results in increasing sea levels which are added to by waters from the melting polar ice caps. Rising global temperatures would trigger this chain of events without deforestation but the process would be slower.

Ocean acidification

Increased carbon dioxide in the oceans is affecting its pH level, creating ocean acidification. Although the pH level is still above 7 (basic/alkaline), the addition of carbon dioxide is raising the ocean's acidity. Some of the carbon dioxide reacts with water to create carbonic acid and compromises the environments of numerous sea creatures and vegetation.

For the last 250 million years the Earth's oceans have maintained a relatively stable pH level of around 8.2. This slightly *basic* level has been challenged over the last 200 years, with the average dropping to 8.1 at present. This reduction corresponds closely to increasing carbon dioxide levels generated by humans since the start of the Industrial Revolution and represents a 30 per cent increase in acidity (the pH scale is logarithmic and the percentage represents this).

Acidification and its impact on marine organisms

Most acidification takes place near the surface and has been shown to inhibit coral and shell growth as well as cause reproductive disorders in some fish. Recent research shows that water chemistry is being altered and the life cycles of marine creatures disrupted, particularly those at the bottom of the food chain. The marine food chain is made up of complex webs and the removal of even one element may result in catastrophic change. The knock-on effect is that marine creatures further up the food chain experience a dwindling food supply and their survival is also put at risk (Figure 3.38 on page 116).

The greatest effect of these changes is being observed in shallower waters where most ocean food resources develop. At present more than 1 billion people worldwide receive their primary source of protein from fish and other marine creatures, and many economies, jobs and communities rely on ocean resources. Changes to the availability of this primary source of protein would necessitate the finding of additional food sources, jobs and even space for these people if they were forced to move to survive. It would not be difficult to imagine these conditions leading to famine, drastic reductions in quality of life, poverty and even conflict.

Pteropods
(planktonic snails)

Limacina inflata

Form the base of many oceanic food webs.

Effects of increasing ocean acidification on pteropods
Extreme scarring and pitting observed on thin outer shells.
Reduced ability to create shells.
Reduced ability to manoeuvre or swim.
Large reduction in numbers.
Observed reduction in numbers happening much earlier than predicted.
Predicted extinction due to speed of acidification giving little time to adapt.

Examples of creatures that are part of food webs with pteropods at their base
Squid
Shrimp
Salmon
Herring
Mackerel
Walleye pollock
Seals
Dolphins
Sharks
Whales
Birds (e.g. herring gulls, cormorants, eagles)
Otters
Bears
Humans
Possible effects?

▲ **Figure 3.38** Pteropods, acidification and the food chain

Oceanographers, concerned at observed changes in marine organisms' behaviours and responses, believe that the 7.6 pH level predicted for the end of this century will have a catastrophic effect on plankton, corals, urchins, oysters, shrimps, lobsters, squid and some fish species. In the case of corals, their demise would put at risk the estimated 1 million species that inhabit coral reefs. The threat provided by increased acidification is intensified for species who also rely on present-day water temperatures for survival.

There will be species that survive and even flourish in the increased temperature and acidified conditions of the oceans. Some observations have shown that jellyfish, some sea urchins, small protein-consuming arthropods, some seaweeds and toxic algae are already adapting and thriving in the changing conditions. Other creatures will benefit as their competitors are either reduced in numbers, have their abilities impaired or become extinct. However, there is overwhelming agreement that many species will not be able to cope or adapt quickly enough and the overall result will be a less diversified oceanic environment.

Oceans are the planet's largest carbon sink and it was formerly believed that their enormity would help to prevent extreme levels of climate change

Research opportunity

Around 252 million years ago the oceans experienced similar acidification. This was at the time of the Permian–Triassic extinction event. It is believed that approximately 90 per cent of marine species became extinct at that time. For a deeper understanding of the past, the present and possible future outcomes of ocean acidification, you will find it helpful to embark on some research. To start with, try an internet search using the following keywords and phrases:

- Permian–Triassic extinction event
- P–Tr event
- oceanic extinctions.

through their ability to absorb large amounts of carbon dioxide (Figure 3.39 on page 117). This absorption rate presently runs at around 20 million tonnes a day, or almost a third of anthropogenic carbon dioxide emissions. It is now believed that the oceans' abilities to act as a sink are becoming compromised due to a number of factors.

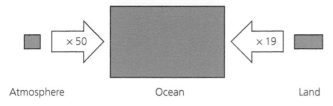

Atmosphere   Ocean   Land

▲ **Figure 3.39** Natural carbon sink ratios

Increasing oceanic temperature

Increasing oceanic temperature has a limiting effect on the oceans' ability to absorb carbon dioxide. Carbon dioxide will dissolve more easily in oceanic waters with lower temperatures than those with higher temperatures. As a result, the increase of oceanic temperature being observed at present means that the oceans' ability to act as a sink is reduced.

To allow carbon dioxide to move from the atmosphere into the oceans, the surface waters must not be saturated with the gas. Microscopic algae called **phytoplankton** help to regulate the levels of carbon dioxide through photosynthesis. Their actions are often compared to that of forests in terms of their ability to remove carbon dioxide from the upper layers of the ocean. In their life cycle they remove the carbon dioxide from the upper layers and move it lower, either releasing it into cooler, deeper waters or to be trapped in the sediments on the ocean bed. Phytoplankton levels have decreased by 40 per cent over the last century, the majority of this being since the 1950s. The decrease in phytoplankton is linked to the increase in ocean temperature. Phytoplankton flourish where colder, nutrient-rich waters well up from deep in the ocean. Scientists believe that increasing temperatures within the upper levels may be preventing deep oceanic upwelling from reaching the surface and may be forcing phytoplankton to starve or operate at lower, less supportive levels.

Reduction in phytoplankton numbers thus reduces the ability of the ocean to absorb carbon dioxide, so leaving it in the atmosphere, and reduces the movement of carbon deeper into the ocean for storage. Scientists see the actions of phytoplankton as a significant contributor to the global carbon cycle and with wide-scale reductions in their numbers it is believed that the ocean sinks are becoming much closer to full saturation. Without the ocean sink, or with a much less effective sink, atmospheric carbon dioxide levels will increase massively and climate change will be stimulated even further and at a quicker rate.

Thermohaline circulation

A change in atmospheric and ocean temperatures also raises the chances of shutting down global **thermohaline circulation** (Figure 3.40 on page 118).This circulatory pattern of water is the major oceanic system that transports heat around the globe. Driven by the different densities of water and their salt content, this 'conveyor belt' has great influence on the Earth's climate. Its warmer waters travel near the surface from mid and low latitudes towards the poles where they moderate the effects of less intense incoming solar radiation and maintain higher temperatures. The **Gulf Stream** and the North Atlantic Drift form part of this circulation. For the North Atlantic and north-western Europe this has a dual effect by limiting the amount of sea ice being produced (and therefore reducing the possible additional cooling effect of increasing albedo) and keeping the area around 9°C above the average for that latitude.

 Task

1 What is the *greenhouse effect* and what is the difference between this and the *enhanced greenhouse effect* when referring to global warming?
2 Outline the *natural* sources of greenhouse gases.
3 Outline the benefits of the naturally occurring greenhouse effect.
4 Look at Figures 3.35 and 3.36 (page 113) and compare the pattern of carbon dioxide emissions with that of average global temperature.
5 What are natural carbon sinks? Give examples.
6 Explain how ocean acidification can affect:
   a) food chains
   b) economies and quality of life.

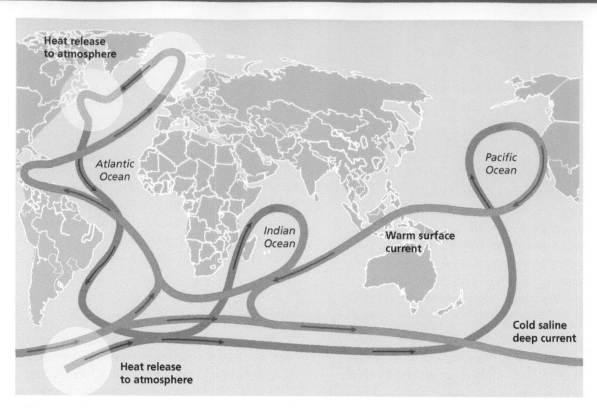

Heat release
to atmosphere

Atlantic
Ocean

Pacific
Ocean

Indian
Ocean

Warm surface
current

Cold saline
deep current

Heat release
to atmosphere

▲ **Figure 3.40** Global thermohaline circulation

A critical part of the thermohaline circulation occurs when warmer surface waters travel poleward and meet colder, denser and more saline waters from the polar regions. In short, these denser waters sink down to the floor of the ocean and move towards lower latitudes. The warmer water moving polewards is cooled and sinks to join the submarine current. The heat released by this cooling warms the surrounding atmosphere. Sinking of these waters powers the circulatory process of the closed system. The cooler waters will eventually mix with warmer waters and rise to the surface in either the Indian Ocean or the Pacific to become part of the warm moving current. Not only does this circulation transport heat but as it sinks, it carries oxygenated water and dissolved surface level carbon dioxide deep into the oceans. The oxygenated water boosts deep ocean environments while the carbon may be withdrawn from the atmosphere for hundreds of years.

Disruption of the thermohaline circulation has the potential to vastly alter global climate. The Atlantic Ocean is the only place where the system transports heat across the equator and to the north. It is the meeting point in the higher latitudes where these waters are forced downwards that drives thermohaline circulation. If this element slows or shuts down, the same will happen to the rest of the global ocean conveyor belt.

Scientists are concerned that increased temperatures are producing large-scale melting of Arctic sea ice (estimating that the Arctic Ocean may be ice free during summers as soon as the 2050s) and Greenland's ice sheet (for example, Greenland's volume of ice is being reduced at a rate of 375 km^3 per year) (Figures 3.42 and 3.43 on page 120). Added to this is increased rainfall and run-off from glaciers and rivers. The release of large volumes of cold, fresh water into the upper layers of the ocean at high latitudes has the potential to disturb or, in extreme conditions, shut down this global system. Studies have reported an estimated 30–40 per cent reduction in the strength of thermohaline circulation in the North Atlantic.

Research opportunity

This is a good point to find out or remind yourself about the thermohaline circulation system. You may already have covered this topic in the Physical Environments: Atmosphere section; if so, take time to go over your notes. If you are not familiar with thermohaline circulation, some research would be a good idea. To start with, try an internet search using the following keywords and phrases:

- thermohaline circulation ● global ocean conveyor
- global ocean conveyor belt.

Previously only thought to affect climate and weather conditions in the Pacific region, it is now believed that El Niño and La Niña have a global influence. In Africa dry conditions are experienced from December to February with higher rainfall from March to May while El Niño is the dominant force in the Pacific. The opposite is found when La Niña is dominant.

Pacific 'neutral' conditions:

- Trade winds blow east to west
- Trade winds bring moist warm air and warm surface water to the west
- Heavy rains through thunderstorms in the west due to convection and moisture
- Drier air returned eastward at high altitude
- Central and eastern Pacific cool
- Cooler waters bring nutrients to the surface in the east and fish flourish

El Niño and La Niña are regular and natural events within the climate system. Episodes usually last between nine months and a year to 2 years (although prolonged events may remain for a number of years). Both are regular events usually reappearing after 2 to 7 years. Recently the regularity and intensity of El Niño has become the most dominant factor. Climatologists believe that this is linked to the increased heating of the Pacific Ocean due to global warming. This could have a great effect on those who rely on the pattern of rainfall or fish stocks to survive. This particularly affects the less developed nations in the regions influenced by the events.

Both El Niño and La Niña are formed due to variations in air pressure in the tropical western Pacific and temperatures in the central and eastern Pacific.

El Niño is the *warm phase*

Formed due to warming of the central and eastern tropical Pacific

Trade winds weaken or reverse

Convection moves towards central Pacific

El Niño the **Little Boy** or **Christ Child** (as it traditionally arrived in South America around Christmas)

Less upwelling of nutrients in the east reduces amount of fish

Warmer than normal temperatures to the eastern Pacific

Much less precipitation in e.g. eastern Australia

Increased convection over eastern Pacific and e.g. Peru

La Niña is the *cold* or *cool phase*

Increased upwelling of nutrients in the east increases amount of fish

Cooler than normal temperatures to the eastern Pacific

Reduced rainfall over e.g. western South America

Increased precipitation in e.g. western, south-western and central Australia

La Niña the **Little Girl**

Formed due to cooling in the eastern tropical Pacific

Trade winds strengthen as they move from east to west

Convection moves, spreads and intensifies further westwards across the East Indies and into the Indian Ocean

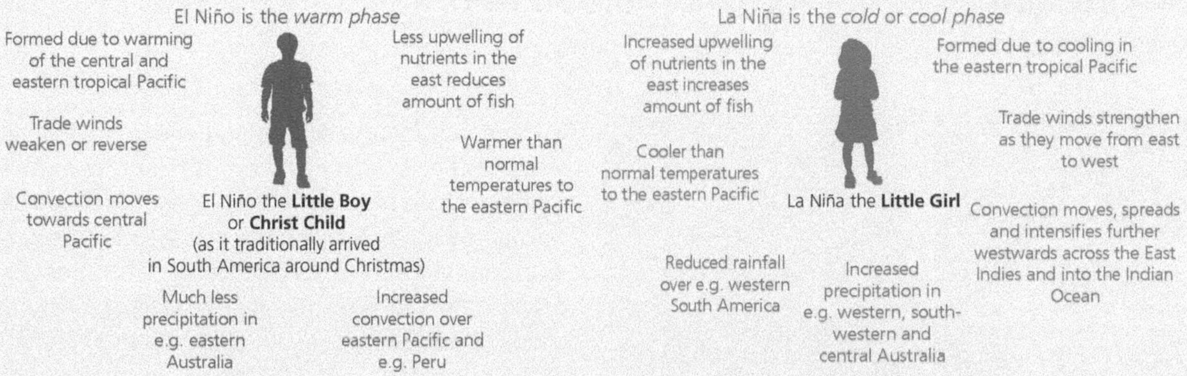

El Niño and La Niña events are a natural part of the global climate system.

▲ **Figure 3.41** El Niño and La Niña

In the Physical Environments: Atmosphere section you should have already learned about El Niño and La Niña. Alterations in their regularity and effects on conditions around the world are being linked to global warming. Take some time to look back over your notes, study Figure 3.41 and use your research skills to find out more.

▲ **Figure 3.42** Meltwater rushes through a channel on the Greenland ice sheet

 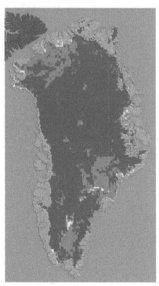

▲ **Figure 3.43** Satellite photographs of Greenland ice sheet melt. The darker areas show increasing areas of ice melt during the summer melt season

With the Arctic Ocean becoming less salty and increasing in temperature, it is predicted that the sinking mechanisms that assist in driving the current could be interrupted. The less salty, less dense water will not sink. Some scientists believe that the current would either switch off, move further south or stutter ineffectively. The consequences for north-western Europe could include (in the worst-case scenario) a drop in temperatures of at least 9°C, altered precipitation levels and extreme winters. One model predicts a cooling of 12°C to the north of Norway, with Scandinavia and the British Isles cooling by up to 4°C on average, while central Europe would experience warming. In addition to this, precipitation across the whole of western Europe would be reduced by around 20 per cent and that shutdown of the thermohaline conveyor would add 25 cm to the height of sea level.

Research opportunity

This is not the first time in the Earth's history that ice meltwaters may have had a climate-changing effect and even shut off thermohaline circulation. You may wish to research Lake Agassiz in North America.

Due to the complexity of climate systems, climatologists have a variety of viewpoints on the effects on thermohaline circulation. At present the majority do not believe that the system will fully switch off but that it could be altered enough to interfere with climate patterns all around the world.

Globally, any change in thermohaline circulation would alter heat and moisture distribution, local weather and climate, as well as oceanic conditions such as oxygen and nutrient transfer between different depths. Without the same level of heat transfer, the land and oceans nearer the equator/thermal equator are expected to increase in temperature. The carbon cycle would also be disrupted, with less carbon dioxide being removed from the upper layers or stored in the deep oceans. Again more carbon dioxide would remain in the atmosphere, assisting in continuing atmospheric temperature increase.

Research opportunity

At present, global ice melt is resulting in around a 2 mm rise in sea level each year. What would happen if large areas of ice totally melted? Use the information in Table 3.3 to work out how the sea-level rises predicted would affect your local area, country and other areas around the world. What would be the consequences for people?

▼ **Table 3.3** Impact of predicted sea-level rises

Location	Volume of ice (km³)	Predicted sea-level rise (m)
Antarctic ice sheets	29,528,300	74.0
Greenland	2,850,000	7.0
All other ice caps, ice fields and valley glaciers	180,000	0.5
Total	32,558,300	81.5

Contribution of other greenhouse gases to global warming

Most of this section has focused on increasing levels of carbon dioxide due to its effectiveness in absorbing/trapping long-wave radiation as it attempts to escape from the atmosphere (it is responsible for around 75 per cent of increased global warming due to anthropogenic greenhouse gas emissions). Carbon dioxide is also present in the atmosphere in relatively large amounts when compared to the other greenhouse gases and more of it is produced by human activity than the rest. However, this should not be a reason to ignore the contribution of other greenhouse gases to global warming.

Many of these gases have a much higher heating potential than carbon dioxide, but like any other facts and figures they need to be interpreted carefully. An example would be methane, with the potential to stimulate 86 times more heating than carbon dioxide over 20 years. However, methane only stays in the atmosphere for around 12 years while 65–80 per cent of carbon dioxide will remain in the atmosphere for up to 200 years. This means that after around 60 to 80 years the impact of the two gases is about the same. What is more concerning, noting the potential for warming from methane, is that concentrations of the gas have more than doubled in the last 200 years.

The thawing permafrost in the Arctic is resulting in the release of large quantities of methane gas trapped within the soils.

Methane

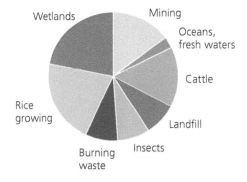

▲ **Figure 3.44** Where greenhouse gases come from: methane

Chlorofluorocarbons (CFCs) are greenhouse gases that also deplete the ozone layer. This depletion allows harmful solar radiation to reach the surface. Although banned or restricted since 1987, these still have an effect and will remain in the atmosphere for up to 400 years.

▼ **Table 3.4** Global warming potential (GWP) of human-produced greenhouse gases relative to carbon dioxide

Selected greenhouse gases	Lifetime (years)	Global warming potential* (relative to carbon dioxide)	
		20 years	100 years
Methane	12.4	86	34
HFC-134a (hydrofluoro-carbon)	13.4	3790	1550
CFC-11 (chlorofluoro-carbon)	45.0	7020	5350
Nitrous oxide (N$_2$O)	121.0	268	298
Sulphur hexafluoride (SF$_6$)	3200.0	16,300	23,900
Carbon tetrafluoride (CF$_4$)	50,000.0	4950	7350

*Global warming potential (GWP) is the amount of heat a greenhouse gas would retain within the atmosphere compared to a similar mass of carbon dioxide. In Table 3.4, the GWP columns show how much more heat the selected greenhouse gas would trap over 20 years and 100 years relative to the same level of carbon dioxide. This gives an idea of the potency of these gases.

The other human-generated greenhouse gases come from a variety of sources, as shown in Table 3.5.

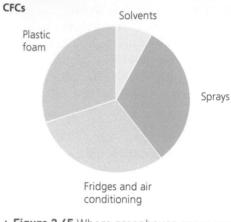

CFCs

▲ **Figure 3.45** Where greenhouse gases come from: CFCs

Research opportunity

The actions of CFCs have created real concerns due to their threats to atmospheric composition, climate change and health. Do some research on their effects and especially how they could seriously affect plant and animal life (including humans). The following keywords and phrases will help you to start your investigation:

- chlorofluorocarbons   • CFCs.
- hole in the ozone layer
- CFCs and skin cancer

▼ **Table 3.5** Major human-generated greenhouse gases

Major human-generated greenhouse gases	Accountability for warming impact from human-generated greenhouse gas emissions	Examples of human-related sources of gas
Carbon dioxide (CO$_2$) (also found naturally)	75%	Burning of fossil fuels for power, heating and transport
Methane (CH$_4$) (also found naturally)	14%	Agriculture (e.g. livestock, rice fields), organic waste in landfill sites, fossil fuel extraction
Nitrous oxide (N$_2$O) (also found naturally)	8%	Agriculture (e.g. livestock waste, nitrogen-based fertilisers) and industrial processes
Fluorinated gases (man-made only)	1%	Industrial processes, solvents, refrigeration, fire suppressants, cosmetics, medicines, electronics, foams, packaging, aerosol propellants
	2% others	

In this complex field of climate change, not all human additions to the atmosphere have a straightforward role or even one that leads to global warming. Black carbon (BC) can have a warming effect and a cooling effect. At higher altitudes it absorbs heat from below and reradiates it to warm the atmosphere. Falling in high enough quantities on ice or snow, black carbon can reduce the albedo effect and also assist warming. However, at lower altitudes it can encourage or stabilise cloud growth and block sunlight and so reduce temperatures. Unlike carbon dioxide, black carbon particles are quickly washed from the atmosphere in as little as seven days.

Other aerosol particles (such as sulphate from fossil fuel combustion) can also help to directly block out incoming solar radiation or act as cloud condensation nuclei to reduce the energy received at the surface. As much as 8–9 per cent of solar energy can be blocked above densely populated areas emitting such aerosols. Again, these aerosols quickly leave the atmosphere.

Changes in land use can also have a cooling effect, where, for example, deforestation can increase the albedo of the area. For all of this, the resulting increase in albedo is well offset by the greenhouse effect of additional carbon dioxide content due to deforestation.

In all, it is believed that the cooling effects of human activities have assisted in slowing the rate of global warming by around 30 per cent, but this still leaves global warming increasing sharply.

Summary

Before the Industrial Revolution, which began around 1750, human activity on the planet had a minimal effect on climate and anything that it did was mostly at local level. The Industrial Revolution and the increasing use of fossil fuels stimulated rapid, global population growth and the need to provide goods and food for people's needs. Our use of machinery, fertilisers, powered transport and energy has grown so much that human interaction with the climate can influence and force change. The main anthropogenic influence has been through emissions altering atmospheric chemistry and, either directly or indirectly through feedback mechanisms, forcing climate and environmental change.

 Task

1 What is global thermohaline circulation and how does it help to moderate the Earth's climate?
2 How could a shutdown or interruption of the global thermohaline conveyor belt affect someone living in Scotland?
3 Create a graph, diagram or visual display to show how the melting of all areas of snow and ice would change sea levels. Be sure to:
   a) make it accurate – use figures and create a scale
   b) show where the water would come from
   c) highlight key facts.
4 Name the major human sources for each of the following:
   a) carbon dioxide
   b) methane
   c) CFCs.
5 Human activities have encouraged the increased output of methane, nitrous oxides and CFCs. Why might these gases be potentially more worrying than carbon dioxide?

Are anthropogenic drivers to blame for global warming?

'Climate change is one of the defining issues of our time. It is now more certain than ever, based on many lines of evidence, that humans are changing Earth's climate. The atmosphere and oceans have warmed, accompanied by sea-level rise, a strong decline in Arctic sea ice, and other climate-related changes ... Much of this warming has occurred in the last four decades. Detailed analyses have shown that the warming during this period is mainly a result of the increased concentrations of CO_2 and other greenhouse gases.'

Source 'Anthropogenic Cause of Global Warming', Royal Society https://royalsociety.org/policy/projects/climate-evidence-causes/

So, why have 97 per cent of climatologists, the world's governments and so many environmentalists become convinced that humans are to blame for global warming? The answer is that so much of the evidence points that way, data and analysis is much improved, and the models created to indicate what changes to expect by human interference are showing a high correlation with what is actually happening in the world. The science is not exact/perfect and there are still some criticisms but the majority of evidence and analysis outweighs that which is against anthropogenic influence.

If Earth had human characteristics and went to seek medical help because its 'climate just did not seem right', what would the doctor do? First, listen to the patient to find out the symptoms, ask questions and take its temperature, blood samples and other tests. Maybe seek expert advice and further specialised investigation. This is done to get the best evidence to define the symptoms and to gain the evidence that would allow the illness to be analysed. Let's look at an excerpt of Earth's medical records after all of this has been done:

- Overall temperature has risen very quickly and is much higher than would be expected.

- Land, ocean and atmospheric temperatures have all shown the same pattern of increase.
- Sea and ocean levels are rising and the rate is increasing.
- Atmospheric chemistry has changed rapidly and includes increasing amounts of greenhouse gases that are accumulating in greater quantities.
- Ice sheets, glaciers and permafrost are melting at a rapidly increasing rate.
- Surface alterations, such as dark patches where there were formerly snowfields or ice.
- Ocean chemistry is altering, with increasing carbon content and acidification.
- Organisms within oceans showing signs of stress, failure to thrive and declining numbers.
- Evidence of early signs of problems or issues in thermohaline circulation.
- Some signs of upset in normal climate relationships, such as increased intensity of storms.

Our medical expert now has to rule out some of the natural influences (physical drivers) on the Earth's climate system – see Table 3.6.

▼ **Table 3.6** Physical drivers on the Earth's climate system

Physical drivers	Comments
Energy emission by the Sun	This has remained relatively constant and has followed its expected 11-year cycle. This has been accurately measured since the 1980s. During that time global temperature has risen but figures still show the effects of the 11-year cycles.
	Conclusion: this is not causing the general pattern of increasing temperature. Additional heating by the Sun would have been expected to have heated all levels of the atmosphere at the existing ratios. This has not happened.
Variations in the Earth's orbit and movement	Eccentricity, axial precession and axial obliquity changes take place over very long periods of time and the changes are slight and imperceptible in human terms and timescales. There has been no vast or rapid change in these that could account for global warming over 200 years or at the advanced rate noted since 1980.
	Conclusion: this is not causing the general pattern of increasing temperature.
Meteorite impact	Meteorites of the scale of the one that impacted at Chicxulub (65 million years ago) or any that have global climate change potential happen very rarely: every 100 million years.
	Conclusion: no such event has happened.
Volcanic activity	Volcanic activity has a mostly cooling influence on climate. Aerosols emitted reflect incoming solar radiation. Their effect is relatively minor, with the largest eruptions in the last 200 years only resulting in cooling for a few years until the aerosols have left the atmosphere. Even recent large events such as Mount Pinatubo (1991) have followed this pattern.
	Conclusion: this is not causing the general pattern of increasing temperature.
Plate tectonic movement	Plate tectonic movement, including continental drift, mountain building and interference with oceanic circulation, takes place over vast periods of time, resulting in changes that are slight and imperceptible in human terms and timescales. There has been no rapid acceleration in these processes.
	Conclusion: this is not causing the general pattern of increasing temperature.

All of these seem to be acting in the normal way and should not have caused any great changes. It is time to examine whether any recent changes in the Earth's lifestyle and behaviour could be responsible. It is at this point that human influences would be analysed, for example:

- increased quantities of greenhouse gases being injected into the atmosphere
- changing surface (deforestation, urbanisation) and its qualities.

A good doctor would also consider possible knock-on effects from these and feedback influences and look for patterns that would suggest that these lifestyle changes in behaviour were to blame. What patterns would the doctor find?

If all of the physical drivers that could influence short-term climate effects are taken into consideration, scientific analysis and climate modelling shows that the Earth should have experienced a decrease in average global temperatures over the last 200 years or so. Only when the influence of human-generated greenhouse gases within the atmosphere is added do the models match the upward trend actually observed as occurring (Figure 3.46). The models also suggest that if it were not for the mitigating effects of natural drivers, average global temperatures would have increased to a much higher figure.

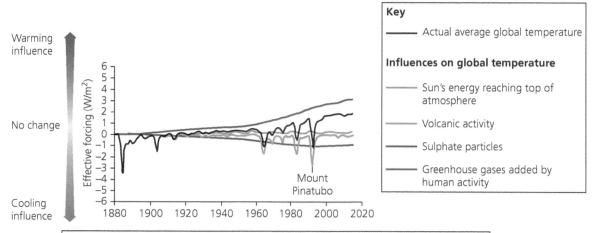

The combined effect of the Sun's energy, volcanic activity and sulphate particles would be to create a reduction in average global temperatures. Only by adding the influence of green-house gases from human activity does the cumulative effect match the actual rise in average global temperature.

▲ **Figure 3.46** Influences on global temperature

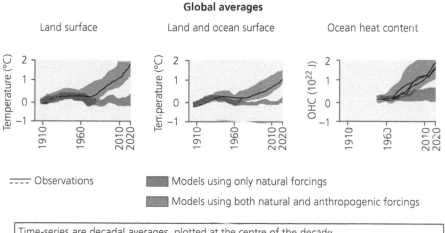

Time-series are decadal averages, plotted at the centre of the decade.
Observations are dashed lines if the spatial coverage of areas being examined is below 50%.
Data from multiple models used.
Shaded bands indicate range from 5% to 95% confidence levels of accuracy.

◄ **Figure 3.47** Land and ocean temperature measurements compared with models of climate change

Not only do the observed measurements and models match for global atmospheric temperatures but they also do so for the land and oceans (Figure 3.47 on page 125).

It would seem to be more than likely that this evidence would convince a doctor that the major cause of the patient's discomfort was anthropogenic – human interference with the normal systems. Perhaps the condition would be labelled 'global warming syndrome' or even 'climate system infection'. Whatever the doctor decides, possible consequences for the Earth would have to be considered along with possible treatments or cures. This is the position in which international agencies, governments, scientists and individuals on planet Earth find themselves at the present time.

Task

Evaluate the influence of human activity on recent climate change and explain why it is the most likely cause of global warming.

3.4 Global and local effects of global climate change

So, what are the prospects for our planet and its climate? Due to the complexity of the climate system, at present we are not completely sure. What we do know is that the computer models now being used are becoming more and more accurate and the suggestions more and more plausible.

Global effects

The Intergovernmental Panel on Climate Change (IPCC), national scientific institutions and large numbers of climatologists have come to a *relative* consensus of what should be expected. Table 3.7 shows the predicted changes for the next century.

▼ **Table 3.7** Examples of predicted climate change in the next 100 years

Air temperature	Storms	Disease
Increase in average temperature of 1.4–5.8°C by 2100. Lower level could only be achieved by a massive reduction in carbon dioxide emissions. The IPCC suggests that the actual figure will be nearer to 3°C. Temperature will not increase equally across the planet. Higher latitudes are expected to experience the largest rises in temperature, e.g. in the Arctic and sub-Arctic areas by 10–18°C.	Increased frequency powered by increased atmospheric and ocean temperatures. These conditions will also stimulate and intensify the water cycle, with increased evaporation, increasing rainstorms, leading to flooding and the like. Tropical hurricanes, cyclones in general and tornadoes to increase in frequency and severity. These may also interlink with intense periods of drought. Many of the cyclonic storms will alter their intensity and geographical location due to alterations of oceanic circulation, temperature and wind patterns. Increased severity of monsoons, along with alteration in geographical area served by the monsoon. Increasing storm activity and intensity for the mid-latitudes, including the British Isles, especially during winter.	Changes in climate and patterns of weather could play a large role in the alteration of locations of diseases. Areas where specific diseases are endemic could change or enlarge greatly. Specific diseases need specific environments and if the conditions are changed to allow for these to survive or spread, the threat increases. Malaria, dengue fever and yellow fever are already spreading as warmer areas expand. With increased temperatures at altitude, some more traditionally lowland diseases are also moving into areas at higher altitudes. Diseases which need insects to act as vectors are seen to be moving location. The increased temperatures and precipitation at higher latitudes are expected to see the return of diseases such as malaria to western Europe from where it was eradicated in 1975.

▼ **Table 3.7** *(Continued)* Examples of predicted climate change in the next 100 years

Air temperature	Storms	Disease
	Some consequences: <ul><li>Increased coastal erosion and flooding</li><li>Increased costs to maintain coastal defences or to repair damage</li><li>Developing countries are not prepared or are less able to cope with storms and damage to coastal defences, as well as difficulties for people after storms</li><li>Increasing of human, infrastructure, environmental pressures when linked with predicted sea-level rise</li></ul>	In the British Isles malaria was once endemic but had slowly been eradicated by the 1950s. It is now projected that by 2050–75 conditions will have changed enough to allow the parasite that causes the illness to thrive. A mosquito that can transmit the disease already exists in Britain (*Anopheles atroparvus*). If the parasite is reintroduced and the damp warm conditions needed for the mosquitoes to multiply in great numbers arrive as predicted, malaria could become a threat. Areas that are predicted to be most at risk from the reintroduction of malaria in the British Isles are the Fens, the Thames estuary, south-east Kent, the Somerset Levels, Holderness, the wetlands of Ireland and the coastal areas of the Firth of Forth. It has been predicted that there could be additional illnesses and disease due to increased temperatures, allowing bacteria to form and grow within our food. These may be bacteria newly introduced to areas. With the increase in temperatures, food would rot much quicker. Food storage and handling would have to be improved.
Precipitation	**Sea level**	**Habitats and species**
There will be a global increase in precipitation but this will not be evenly distributed geographically. In general drier areas will get drier, and wetter areas wetter, with some anomalies: <ul><li>Northern latitudes (including the British Isles) will experience an increase in precipitation, especially during winter</li><li>North America, South America, southern Europe, Africa, Middle East, Central Asia will experience much lower rainfall and more frequent droughts</li></ul>	There is a predicted increase caused by expansion of ocean waters due to increased temperatures and melting of global ice and snow. There is a potential rise of around 80 m solely from the melting of global ice and snow. Models have shown that this will not have happened within the next hundred years but that the following figures are possible: <ul><li>Best-case increase of around 10–25 cm</li><li>Worst-case increase of around 3.6 m</li><li>Not only will this cause coastal or river valley flooding, it will also cause salt water to seep into agricultural land and water supplies. Developing countries will be less prepared or less able to cope with the changes</li></ul> Examples of places at risk are the Mekong Delta (Vietnam), the Nile Delta (Egypt) and the Ganges Delta (Bangladesh). These are areas of high population and important for agricultural production and food supply.	Human habitats will come under stress. There will be movements of people away from areas where life becomes increasingly more difficult (due to lack of fresh water, lack of food supply and reduction of land area available). Refugees will be seeking new land to live on and people will suffer from starvation and death. All of this will put strain on the world's economies, standards of living and quality of life. It is possible that more conflicts will arise as individuals and groups fight for control of resources. Similar to human beings, all other living creatures will experience changes in habitat. Ecosystems will shift. Already there is evidence of changing spatial locations of some animals, insects and plants. There has been a northerly movement for those where previously the cooler temperatures were not suitable. Others have moved to

▼ **Table 3.7** *(Continued)* Examples of predicted climate change in the next 100 years

Precipitation	Sea level	Habitats and species
<ul><li>Prairies (Canada and USA), Pampas (Argentina) and the Steppes (Russia), the world's centres of grain production, will experience large-scale reduction</li><li>Areas of marginal farming capability, such as southern Europe and the Sahel of Africa, could experience desertification or extended desertification</li><li>Australia, Central America and sub-Saharan Africa could see a decline in precipitation of 30% by as early as 2050</li></ul>Water shortages to affect hundreds of millions, with around 250–260 million exposed to increased water stress by as early as 2025. This figure is expected to rise with a continuing increase of temperatures and reduction in precipitation in vulnerable areas:<ul><li>16% of world population to suffer from water shortages</li><li>25% reduction in economic output globally</li></ul>Some consequences:<ul><li>Decrease in world cereal production of 400 million tonnes predicted</li><li>Global food shortages</li><li>Major reduction in animal feed</li><li>Massive increase in famine and geographical area where famine is experienced</li><li>25% reduction in economic output globally</li><li>Increased number of refugees</li></ul>	<ul><li>Mekong Delta: known as the 'rice bowl' of Vietnam. It grows 47% of the national total of cereals. It is predicted that by 2030 millions of tonnes of agricultural produce will be lost from this area and many of its regions will be flooded. Much of the soil is beginning to experience an increase in salinity</li><li>Nile Delta: a rich agricultural area with a population of around 80 million. 40 million are believed to be at threat from climate change induced sea-level rise. It is predicted that by 2025 200 km² of land will be lost. It is estimated that by 2100 the loss of agricultural land will have created food shortages and approximately 7 million climate refugees</li><li>Ganges Delta: one of the most fertile regions of the world, giving it its nickname 'The Green Delta'. Between 125 and 143 million people live there. It is believed that 300 million people are supported by the agricultural and fishing outposts of the Ganges Delta. It is predicted that by 2050 3 million people will be affected by sea rise and 8% of rice production will have dropped as well as 32% of wheat. By 2100 Bangladesh will have lost around 25% of all of its land area. Not only will millions be affected by the loss of land, space and food, but freshwater reserves will have been vastly undermined</li></ul>Even in the British Isles, the cost of maintaining coastal defences may become cost-prohibitive, resulting in land being abandoned to the sea. Coastal and river valley housing and industry will be affected, as well as agricultural land. Areas such as East Anglia and London will be under great pressure.	higher altitudes as temperatures also warm. Those which survive well in the cooler temperatures have also seen a decline. Arctic creatures such as polar bears have found their traditional hunting areas on the ice packs become limited or dangerous. This has resulted in famine and death and in some cases the movement further south and onto occupied lands. In the oceans food webs are also becoming disrupted, due to temperature change and acidification and there are already species in decline. Where creatures/plants can adapt, move or enjoy improved conditions for their species they will survive but it is predicted that there will be a general decline in biodiversity. There have been some suggestions that the increased carbon dioxide content may have a positive benefit to some plant life. They may thrive and grow to sizes larger than at present.

Case study: Tuvalu

(5°41'S 176°12'E to 10°45'S 179°51'E)

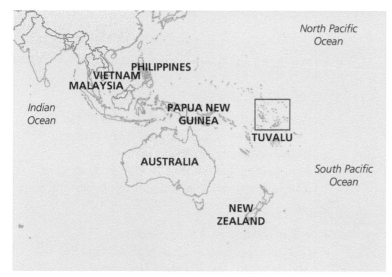

▲ **Figure 3.48** Location of Tuvalu

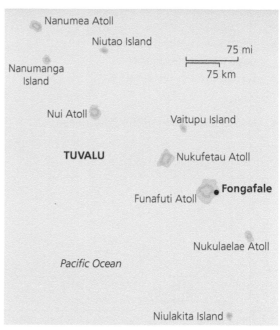

▲ **Figure 3.49** Map of Tuvalu

Tuvalu is a small, 26 km², Polynesian island state in the Pacific Ocean, some 5153 km (3202 miles) from Australia (Figure 3.48). It is the fourth smallest country in the world and a member of the British Commonwealth. The nation consists of six atolls and three reef islands (Figure 3.49). These are called Funafuti, Vaitupu, Niutao, Nanumea, Nanumanga, Nukufetau, Nui, Nukulaelae and Niulakita. The atolls and reefs are made from coral with the highest points only a few metres above sea level and this makes it the country with the second lowest maximum height. The highest point is 4.6 m on the island of Niulakita.

The population of Tuvalu was stated as 11,329 in 2018 but it has a high population density of just over 379 people per square kilometre. Tuvalu has a mixed market subsistence economy but is not self-sufficient in agricultural production.

Tuvalu is very vulnerable to the effects of climate change and in particular rising sea levels and salinisation (page 132).

▲ **Figure 3.50** Tuvalu: low-lying islands and reefs

Projections, problems and responses

Current projections suggest that Tuvalu may become uninhabitable in 50 to 100 years and that the last remaining islands will vanish beneath the Pacific Ocean by 2100. Already some scientists are claiming that Tuvalu will become the first nation to disappear due to the effects of global warming. So what are the Tuvaluan people experiencing now and what is expected in the near future?

Air temperatures

- Increase by up to 4°C by 2100 (Figure 3.51).
- Increase in number and intensity of extreme heat days and warm nights.
- Decline in cooler weather.

Response:

- Not in the hands of the islanders.

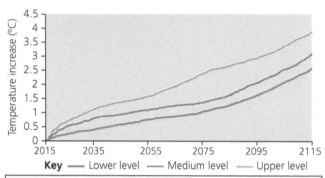

Key —— Lower level —— Medium level —— Upper level

The graph is based on models investigating three scenarios:
- low global carbon dioxide emission rate
- medium global carbon dioxide emission rate
- high global carbon dioxide emission rate.

All of the models showed a projected increase in temperatures. The graph shows the range of results amalgamated from all of the models:
- lower level temperature increase (lowest projection)
- medium level temperature increase (middle projection)
- upper level temperature increase (highest projection).

Note that the model is skewed due to the time intervals selected.

▲ **Figure 3.51** Tuvalu: range of temperature increase projections, 2015–2115

Ocean temperatures

- These will continue to rise. At present, surrounding sea area is experiencing a reduction in numbers and diversity of fish types; this is linked to temperature increase and acidification.
- Expansion of ocean water due to increasing temperatures resulting in flooding incursions, increased coastal erosion and loss of land.

Responses:

- Not in the hands of the islanders.
- Coastline protection on a large scale is not affordable for the Tuvaluan Government (see 'Sea level and erosion' on the right).

Precipitation

- Predicted increase in average annual and seasonal rainfall. Increases during the wet season while dry seasons intensify resulting in drought.

- Increased intensity and frequency of extreme rainfall days.
- At present some inland agricultural areas have reported increased crop failure due to flooding and intensity of rainfall.

Responses:

- Not in the hands of the islanders.

Tropical cyclones

- Projected to reduce in number by 2100 but to increase in wind speed by 2–11 per cent and rainfall intensity of around 20 per cent.
- At present ocean surges flood inland, washing away or damaging homes and removing vital soils.

Responses:

- Not in the hands of the islanders.
- Coastline protection on a large scale is not affordable for the Tuvaluan Government (see 'Sea level and erosion' below).

Sea level and erosion

- Predicted to continue to rise from its present level of 4 mm per year, with a possible overall rise of 4–14 cm by 2030.
- Some models predict that by 2100 the rise may be 32–60 cm (Figure 3.52).
- These rises are extremely concerning as many parts of Tuvalu have an elevation of less than 1 m and the highest point is only 4.6 m.

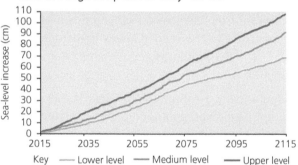

Key —— Lower level —— Medium level —— Upper level

The graph is based on models investigating three scenarios:
- low global carbon dioxide emission rate
- medium global carbon dioxide emission rate
- high global carbon dioxide emission rate.

All of the models showed a projected increase in sea level. The graph shows the range of results amalgamated from all of the models:
- lower sea-level increase (lowest projection)
- medium sea level (middle projection)
- upper sea level (highest projection).

Note that the model is skewed due to the time intervals selected.

▲ **Figure 3.52** Tuvalu: range of sea-level increase projections, 2015–2115

- Already tides are rising higher and causing increased flooding.
- Pools of sea water on land are regular occurrences and for several months a year aircraft have difficulty landing on the flooded airport runway.

▲ **Figure 3.53** Funafuti, showing its international airport

- Encroachment from the sea has claimed over 1 per cent of the land area that was present in the 1950s.
- Te Pukasavilivili, a small island on the edge of Funafuti atoll, disappeared in 1997.
- Shorelines are retreating and the atolls shrinking.
- Sandy beaches that once added some protection against the power of the waves have been eroded and buildings that were once a distance from the shore are now under threat.
- Former burial areas have now become inundated, with some people choosing to exhume bodies and bury them farther inland.
- With sea water seeping farther inland, the roots of coconut trees are rotting, causing them to die. Entire atolls that were previously covered with trees are now left bare and open to further erosion.
- Large areas where people lived or used to grow crops near the shoreline have been worn away, reducing the ability to grow food and destroying many people's livelihoods.

Responses:

- Mangroves are being planted at the coastlines to help reduce erosion (Figure 3.54).

▲ **Figure 3.54** Tuvalu: mangrove plantation

- New houses are being built on 3 m high stilts and some previously constructed homes and businesses are being raised (Figure 3.55). This is not a traditional style of architecture in Tuvalu.

▲ **Figure 3.55** Tuvalu: houses on stilts

Although some local attempts have been made to build walls to provide protection from storm surges and sea-level increase, it would be cost-prohibitive for the government to attempt to provide such protection for all the islands.

The country is a developing nation, with few resources and very little money (the UN classifies Tuvalu as a least developed country (LDC)).

Any scheme to attempt to protect the country's coastlines would cost billions. It is estimated that a temporary sea wall for one small atoll could cost over £110 million (approximately $175 million).

Ocean acidification

- It is predicted that the acidity level of ocean waters surrounding Tuvalu will continue to increase throughout the twenty-first century.
- The increased acidification is expected to impact on the health of the reef ecosystems (reduction in number and diversity of fish and other life forms).
- Coral bleaching, and the crumbling of the islands and reefs made from coral, will make the islands less habitable through reduction in the ocean-provided traditional foods. Already reductions in numbers and types of fish are being reported.
- The reefs of the island of Funafuti have already suffered 80 per cent coral bleaching from increased ocean acidification and temperature.

Responses:

- Not fully in the hands of the islanders.
- Coastline protection on a large scale is not affordable for the Tuvaluan Government.
- Mangroves are being planted at the coastlines to create areas for fish to breed and thrive.
- With Japanese scientific assistance and aid, methods to assist reef restoration are being researched. As corals expel the algae living in their tissues, causing the coral to turn completely white (bleaching), the Japanese project is experimenting with adding foraminifera (a type of single-celled amoeba with a shell) into the bleached coral to replace the algae and reinvigorate the reefs.
- A Funafuti Conservation Area has been created to limit the fishing in the lagoon, so as to conserve and hopefully, in time, increase the depleted fish stocks.

Salinity, agriculture and water supply

- Tuvalu's chief export is dried 'coconut meat' but large areas of land used for palm plantations have become unusable due to sea water flooding or through salt water seeping into the soils and contaminating them (salinisation).
- Sea water is not only affecting the coastlines but is seeping farther and farther inland.
- Along with fish, coconuts and starchy vegetables like breadfruit, pulaka and taro form the traditional diet.

Pulaka is the staple diet of Tuvalu. Its roots are rich in nutrients and it is grown by digging pits into underground fresh water a few metres below the surface.

- The ability to grow pulaka is being reduced as sea water seeps into the pits. Some areas have seen a 75 per cent reduction in the ability to grow pulaka. Fruit-producing trees are also suffering.
- Tuvalu now struggles to provide even at a subsistence level and relies on imported food.
- Tuvalu's islands have no rivers or streams and fresh water on the islands has been deposited by rain. Groundwater is a limited resource and is becoming undrinkable due to sea water contamination of the island's **aquifers**.
- Rainwater harvesting (catchment) is a major source of water, being directed into tanks from roofs for example. Periods of drought are expected to increase, causing shortages of water.

Responses:

- The Food and Agriculture Organization of the United Nations is financing the introduction of salt-resistant banana plants and taro roots to provide a replacement food supply.
- Mangroves are being planted at the coastlines in a double effort (preventing erosion and providing a habitat for fish to help sustain this vital part of the islanders' diet).
- In recent years desalination has become a major source of fresh water. Japan funded the purchase of desalination units and Australia and New Zealand provided temporary units to respond to drought conditions.
- The European Union and Australia also provided water collection tanks. It would have been difficult, if not impossible, for the Tuvaluan Government to afford such support mechanisms and machinery.

Evacuation, refugees and culture

Tuvaluans are very aware of their plight. In under 100 years they may all have to abandon the islands and leave them to their fate beneath the waves. Already parts of the islands have become uninhabited. There is a general wish not to leave but, with global warming offering such a bleak future, it looks inevitable. Even large-scale reductions in greenhouse gas emissions from now on would not appear to be enough to prevent the 'sinking' of these islands. Already the islands' government has been in negotiation with New Zealand and Australia about immigration programmes to allow Tuvaluans to move there, at first in small numbers but with the possibility of larger-scale immigration later. There is a fear that once the people leave the islands their culture will be lost forever. In their own language they have a word – 'fakaalofa' – which is used to describe people who have no land: this means 'deserving of pity'. Many now use this to describe what they will be if they leave the islands.

Task

1. Create a diagram, media information package (for example, PowerPoint presentation/website report) or a wall display to show the possible effects of global warming between the present and 2100.
2. Make a list of the local effects of global warming.
3. Using the case study on Tuvalu and information from the rest of this chapter, answer the following:
   a) Why is Tuvalu at extreme risk from the effects of global warming?
   b) Give two reasons for sea-level rise due to climate change.
   c) Explain the main risks and difficulties faced by the Tuvaluans.
   d) Evaluate the attempts to improve conditions in Tuvalu.
   e) Using all the evidence from graphs, diagrams, information and your own research, evaluate the position the Tuvaluans are in and whether you believe that people will still inhabit Tuvalu by 2100.
   f) Do you think that the term 'fakaalofa' is an effective word to describe the future position of Tuvaluans? Give reasons for your decision.

Research opportunity

Tuvalu is a good example of the possible outcomes of uncontained global warming, but there are other examples. Table 3.7 on page 126 mentioned other areas where there were major concerns (for example, Vietnam, Egypt, Bangladesh, the Amazon Rainforest and the Prairies). Perhaps you would like to find out more about these places or somewhere else in which you are interested. This would provide you with a different example for your exams or your assignment.

Effect of climate change on Scotland

So far very little has been mentioned about Scotland except that we could experience extreme cold if thermohaline circulation was badly disrupted or stopped altogether. If we remove this possibility, what are we likely to experience in Scotland? Figure 3.56 is based on Scottish Government projections as well as those from other environmental agencies. This shows that, in general, Scotland will get warmer but wetter, with summers being more intense but shorter and winters stormier but milder. Although not looking at all aspects of the climate change predicted, it does suggest that environments will change a great deal and affect all our lives.

Reduced snowfall and frost. Results in higher snowline. Ski industry severely affected. In mountainous regions endemic Arctic species ptarmigan, Arctic hare, Arctic-alpine plants lose habitats and edge towards extinction. These may be replaced by species whose habitats spread further north or in altitude. Mild winters stunt growth of Sitka spruce.

Agricultural environment changed. Traditional crops may not be the best for the new warmer climate conditions. Crops previously unable to survive become replacement or better options (e.g. maize). Double cropping of fruits such as strawberries and raspberries. Additional spending may be required on insecticides and also irrigation for drier growing seasons.

Increase in average temperature by 1–3 °C by 2050. Increase in heatwaves and summer drought conditions. More intense but shorter summers. Increase in summer heat-related deaths (young children, older adults, people with medical conditions especially). Milder but stormier winters. Reduction in winter cold-related injuries and deaths. Warmer temperatures increase the concentrations of air and water pollutants, creating unhealthy conditions.

Increase in speed that food/crops can rot either due to increased heat or heat/moisture. Increase in food-related illnesses (higher temperatures may cause increase in cases of salmonella – bacteria grow more rapidly in warm environments and other food poisoning bacteria may also thrive in the new conditions).

Increased rainfall in western Scotland by 25–40% by 2050, mostly during the winter but also including intense thunderstorms in summer. In general rainfall will be more intense. More flooding on river floodplains (especially during winter), increased cost for home insurance, less building space, additional costs of building flood defences. Homes, roads, power stations and other buildings threatened.

Rise in sea level by greater than 40 cm. Extreme coastal weather events increased in regularity and strength. Sea walls will need to be strengthened or raised or lower land abandoned (e.g. Firth of Forth). Homes (and other buildings) and transport networks lost or regularly damaged. Increased coastal erosion.

Northward movement of species of insects due to increased temperature. Possible return of malaria (e.g. Firth of Forth).

Houses (especially older ones) will experience increased rot due to intense rainfall and increased temperatures. Newer materials may have to be found for construction and roads. Sewerage system (especially in cities like Glasgow) is not designed to cope with increased frequency of storms and intensity of rainfall. Large-scale expenditure will be needed. Flooding from sewers could cause additional disease/illnesses.

Increase in sea temperature. Traditional fish species such as cod and haddock forced to move further north. Southern species may reach as far north as Scotland (e.g. tuna).

▲ **Figure 3.56** Scotland: projected climate change 2050–75

Task

1   Discuss the possible effects of global warming in Scotland.
2   Analyse the possible effects of global warming in Scotland and describe how they could affect you in terms of:
   a) home life
   b) health
   c) financial cost
   d) job opportunities
   e) travel.

3.5 Management strategies and their limitations

It has become increasingly obvious that action needs to be taken on both a global and local level if climate change is not to reach a level that is unmanageable and harmful to our environment and even the continued existence of human beings on our planet (Table 3.7 on page 126). So what needs to be done? Figure 3.57 gives some suggestions.

A worldwide approach is vital if we are to ensure:

- reduction of carbon emissions and those of greenhouse gases in general
- research, investment in and utilisation of technologies that will assist in reducing greenhouse gas emissions and those that will assist with adaptation to change
- reductions in forest loss

- assistance to those countries economically less able to reduce emissions or to adapt to change.

With so many individuals, industrialists, investment groups, governments, political systems, scientists, viewpoints and short-term needs to be satisfied, international agreement and action have so far proved difficult to achieve. What is apparent is that there is much more agreement that action needs to be taken.

Problems particularly exist in developing countries and those whose populations are expanding rapidly (for example, China and India) where they need to find ways to improve or simply to maintain their standard of living by using the resources they have available. Some developing countries do not have the finances available to adjust their industries so as to reduce emissions. A view has been established that developed nations must take additional responsibility to financially assist those who would find difficulty in reducing emissions.

Agriculture, forestry, land use	Transport	Industry	Waste	Residential and commercial heating and power	Power generation
• Increase in environmentally sympathetic farming methods to reduce emissions • Increase in locally sustainable methods of farming to reduce transport emissions and the need for chemicals • Reduce use of chemical-based fertiliser (reduces emissions associated with both production and use) • Encourage human diet change (less meat, milk, cheese and butter) to reduce livestock numbers and animal methane emissions • Lower livestock ratio (fewer animals per hectare) • Reduce food waste • Afforestation (large-scale planting of trees to absorb carbon) • Restoration of wetlands, which may have been previously drained, to store more carbon	• Improve vehicle engines to reduce emissions by 25% • Strict enforcement of speed limits (faster speeds result in more emissions) • Increase tax on petrol and diesel • Increase tax on vehicles with high fuel usage and/or high emissions • Lower road tax for cars with smaller engines • Phase out or ban vehicles with petrol or diesel engines • Encourage use of electric vehicles • Walk, cycle or use other non-greenhouse gas-emitting methods of transport for shorter distances • Encourage use of public transport • Use video conferencing in place of business travel	• Better management of processes and the introduction of new technologies to result in low, zero or negative carbon emissions • Use of carbon-negative building materials (wood, carbonated aggregate) • Carbon sequestration (carbon capture and storage, CCS, which entails trapping CO_2 and preventing it from entering the atmosphere) • Strict legally binding controls on the emission of greenhouse gases • Use of cleaner heat sources, such as hydrogen	• Increase domestic, commercial and industrial recycling and reuse • Separate collection of food waste to divert from landfill (food produces methane if buried and allowed to rot) • Strict management of industrial and domestic waste facilities and the application of new technologies to reduce emissions • Banning waste that contains biodegradable carbon from landfill • Capture methane emitted from landfill sites	• Use hydrogen as a replacement for natural gas until renewables become dominant • Improve insulation in homes, commercial and public buildings, reducing demand for power generation • Greater energy efficiency in homes, public and commercial buildings (thermostats set to lower temperatures, more efficient heating systems and electrical appliances) • Use of low-energy light sources • Encourage consumer choice of energy providers that supply energy from renewable sources • Use heat pumps, which use electricity to move heat around, rather than heat sources or air conditioning • Encourage use of washing lines instead of tumble driers	• Develop and apply renewable energy sources (wind, wave and solar power) to replace carbon-based energy production from oil, gas and coal while also reducing the need for nuclear power • Encourage consumers to choose power generated from renewable sources and suppliers to provide power from such sources • Develop energy storage and smart systems which allow for the most efficient use of energy • Develop substantial CCS infrastructure to apply to power-generating plants

▲ **Figure 3.57** Some ideas on how to tackle the problems of global warming

Worldwide agreements

Since the early 1990s, there have been concerted efforts by governments around the world to find agreements on how to tackle climate change. Much movement forward has been made on the identification of the issue, the need for action and the understanding that action needs to be taken quickly.

At the 2015 United Nations Climate Change Conference in Paris (COP21), for the first time almost every country in the world agreed to act to reduce greenhouse gas emissions. The agreements set a target of keeping global temperature increase to below 2°C.

This was a great breakthrough in tackling climate change, with all the major producers of greenhouse gases signing up. Unfortunately, due to political change, the USA (the world's second largest greenhouse gas emitter) later announced its intention to withdraw from the agreement by 2020 and to lift many of the restrictions on carbon-based fuels and resources.

This came as a double blow along with the 2017 announcement that, even if all the signatories of the Paris agreement held to their promises, global temperature would still increase by more than 3°C by 2100.

In 2018 the United Nations' Intergovernmental Panel on Climate Change (IPCC) re-evaluated the effects of the targeted 2°C rise on the planet and realised that such an increase would not be 'safe'. Instead, the IPCC recommended a target of 1.5°C and the need for much tougher and quickly applied restrictions on greenhouse gas emissions, while also adopting other techniques to reduce global warming.

It warned that even a 1.5°C rise would still bring severe consequences and changes to the planet, but these would be much less than with the 2°C limit set in 2015 or the more likely worst case scenario of 3°C.

Although 0.5°C less of an increase may not seem much, this would result in (for example):

- 10 cm less in sea-level rise when compared with a 2°C temperature increase, meaning 10 million fewer people would be under threat from rising sea levels

- 23 per cent fewer people being exposed to extreme heatwaves every five years, when compared with a 2°C temperature increase.

When does climate change become irreversible?

There is a general understanding that reversing climate change may not be possible, but we may be able to halt global temperature increase at a level where the climate and its system is such that human existence on a large scale is able to remain. The United Nations has agreed that an increase in temperature of 1.5°C above that of pre-industrial levels is the **tipping point** for our planet's climate system. The view is that after this point climate change will become irreversible and that any chance that people could stop or mitigate the negative effects will have gone (Figure 3.58 on page 137). It should be remembered that global temperature has already increased by more than 0.5°C since industrialisation and that, without any action, some models suggest there could be another 2–3°C rise within 50 years.

More predictable results would follow the breaching of the 1.5°C increase level:

- 4 billion people would experience water shortages.
- An additional 200 million people would suffer from hunger.
- Crop yields would drop by 30 per cent across the continent of Africa.
- A further 60 million people across Africa would become exposed to malaria.
- Up to 40 per cent of wildlife species would become extinct.

As mentioned earlier, despite all the commitments agreed at the Paris Conference to hold global warming below 2°C, the United Nations recognises that we will not succeed to reach even that target. Many nations have failed to meet the targets that they promised, with some not having implemented **mitigation** processes or having the systems to record, check or enforce their commitment.

Whatever action is taken it will have difficulty in slowing down or stopping global warming. As we have already noted, carbon dioxide levels have been building in our atmosphere for over 300 years and

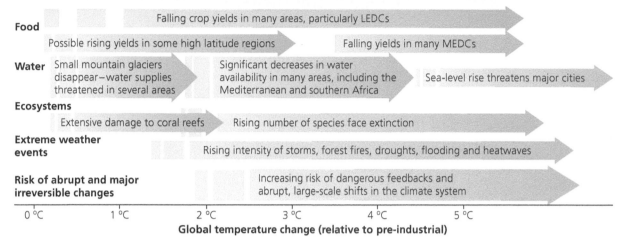

▲ **Figure 3.58** Projected impacts of climate change

the gas remains in the atmosphere. In addition to this, carbon dioxide has an atmospheric lifetime of between 50 and 200 years so, even if all carbon dioxide emissions from human activities stopped immediately, the gases already in the atmosphere would continue to be active for quite a considerable time. The action from these gases could still trigger feedback processes that have the potential to maintain the warming and push it past its tipping point.

Carbon trading

Carbon trading is based on agreements within the Kyoto Protocol (1997). This is a scheme where countries (or businesses) buy and sell carbon permits as part of a programme to reduce carbon emissions. Countries are given limits on the amount of emissions they are allowed and, if they find they need to go above their limit, they may purchase spare permits (**carbon credits**) from countries that have produced less than their quota. Countries, in turn, set quotas for businesses and they too can trade their credits in a similar fashion. This process is known as **cap and trade**.

The intention of carbon trading is to place a monetary value on the emission of greenhouse gases. Rather than just discarded waste, atmospheric emissions become part of the cost of creating and pricing a product. Businesses are then forced to find methods

to comply with the quotas or be fined large amounts. To remain competitive in the marketplace, companies must find and invest in ways of making their processes cost effective.

There have been criticisms of this process, mainly based on whether the system is fair and equal. Some see this process as really benefiting developed countries while reducing the ability of developing countries to improve their quality of life. Developing countries' industries may not have the financial ability to compete for the purchase of carbon credits and are isolated from the benefits of the system. In addition, some of the world's largest polluters, including the USA, did not sign up to the protocol and were seen to have an unfair advantage as they did not have to comply with restrictions and their products could be sold on the international market at a cheaper price as costs were lower.

Other criticisms highlight the potential for fraud. Accusations have been made that in a number of countries, such as Russia, China and India, companies began producing greenhouse gases solely to be given a quota and then reduce emissions so as to sell their carbon credits elsewhere. The need for improved monitoring and verification of the system has been highlighted. It has been suggested that the increased utilisation of monitoring by satellites will assist in the accurate measurement of emissions and even land-use change.

Carbon offsetting

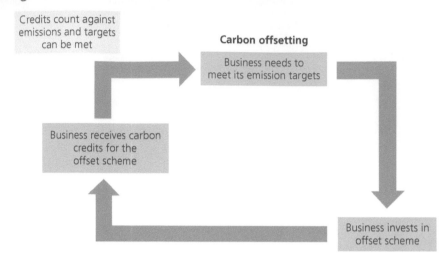

▲ **Figure 3.59** Carbon offsetting

Countries and companies may also invest in **carbon offsets** to compensate for emissions above their quotas and allow them to meet their obligations (Figure 3.59).

These credits can be applied by investing in schemes elsewhere that will remove greenhouse gases in the atmosphere. These schemes include planting of trees, renewable energy, efficiency projects, destruction of pollutants from industry or agriculture and development and use of alternative carbon storage (Table 3.8).

▼ **Table 3.8** Top carbon offset schemes

Renewable energy	Improving energy efficiency
Reason: To provide non-carbon-based sources of power to reduce emissions. By 2020 the Scottish Government intends, through its energy plan, to generate 100% of electricity by renewable sources and 50% of total energy consumption by 2030. **Criticism/limitations:** As offsets these may provide zero carbon energy (although construction or transportation to sites may involve carbon emissions) but are only effective in reducing global warming if the emissions from already traditional methods of power generation are reduced. Wind, wave, tidal and solar power all have environmental criticisms such as disruption to local ecosystems. Hydroelectric power often necessitates the building of large-scale dams which again disrupts local ecosystems but may also include large-scale greenhouse gas emission in, for example, the production of cement.	**Reason:** To reduce the amount of energy being generated and as such lower the amount of fossil fuels burned and greenhouse gases released. **Criticism/limitations:** There is a great deal of support for this type of initiative as, for example, insulation, energy efficient household items and low-energy light bulbs reduce the need for the generation of power. Schemes in developing countries such as providing more efficient stoves and low-energy light bulbs are seen as very positive. Its criticism is that it does not encourage a change in attitude towards power use. People (especially in developed countries where massive power consumption is the norm) could still switch on as many electrical items but expect the technology to assist rather than consciously making use of less power.
Forestry	**Destruction of industrial/agricultural by-products**
Reason: To replace trees being cleared. To restore and enhance the ability of forests to act as carbon sinks. **Criticism/limitations:** Trees take decades to grow to full size and may not have a quick enough effect on the levels of carbon dioxide in the atmosphere. There is no guarantee that the trees will not be cut down later. New forests are not as effective or biologically diverse as the ancient forests (e.g. tropical rainforests) they may be trying to replace. Replanting may harm environments and even displace local people.	**Reason:** To reduce potential gas emission from source or waste products. **Criticism/limitations:** These schemes can provide valuable assistance where the technology and expertise are either not available or unaffordable. Criticism is targeted at companies who did not invest in destruction of their own wastes while paying for offsets.

Investment in schemes that will assist developing world countries, such as helping to provide carbon capture at their industrial sites, is encouraged and helps to fight off criticism that carbon trading is for the benefit of wealthy countries only.

Carbon offsetting has been criticised for letting the wealthy feel better about what they are doing without really changing their ways. Economically developing countries would again be asked to make the changes, and even be the location for the schemes, yet not reach the standards of living and comfort of those who have contributed most to creating the problem.

Some investments are made only on short-term projects that may not have the desired effect in combating wide-scale global warming. Other criticisms include buying into schemes that are not scientifically proven to have the ability to compensate for the emissions actually being made. It has been suggested that a tighter control of the type of schemes and their effectiveness allowed under the system should be built into future protocols. Increased utilisation of monitoring by satellites will, as mentioned, assist in the accurate measurement of emissions and even land-use change related to carbon trading and offsets.

The greatest criticism of carbon offsets is that they should not be used at all. Environmental groups point out that if governments were sincere in their approach to tackling global warming these schemes would be financed anyway, while the emitters should be forced into reducing their emissions.

Carbon taxes

Another method used to reduce carbon dioxide emission is the imposition of **carbon taxes**. These can be used to levy fees on the production, distribution or use of fossil fuels. The taxes are usually based on the amount of carbon emitted by the particular fossil fuel when burned. These taxes are favoured by many climatologists and economists as they are relatively simple to impose and target the emitters and their markets by forcing costs on them. Unlike carbon trading systems, where lengthy periods of time are necessary for rules, monitoring and economic organisation to develop, most governments already have taxation and administrative structures that allow for speedy implementation.

Carbon taxation has a dual target of making those responsible for emissions pay and encouraging them to find methods that will reduce greenhouse gas emissions (for example, reductions in emissions, more efficient production methods or a switch to low or zero carbon energy). All involved in this are incentivised to make changes, both at production and customer level. Supporters of carbon taxes see them as a way to make polluters pay for the societal costs resulting from emissions that are made by them or generated in preparing items or services for them.

Critics of carbon taxation point out that it is often the financially deprived members of society (for example, low-income households) who are affected the most. If costs are allowed to rise due to the imposition of carbon taxes, people with less money find it difficult to purchase products, including electricity, petrol and heat. Policies such as tax credits or tax reductions can be used to mitigate against affecting the less well off. Some countries, such as the Netherlands, use the monies raised from carbon tax to reduce the overall tax burden on their population. To further reduce the effects of carbon taxation, some governments issue fines to companies who pass on the additional costs to their customers.

Quotas and taxation are part of a general movement towards finding more efficient methods of reduction in the use of fossil fuels and alternative ways to dispose of gases (for example, sequestering carbon dioxide in artificially created or reused sinks such as the layers of rock from which North Sea gas has been extracted).

The movement towards low/zero carbon emitting power sources has seen a speedy development in alternatives or an increase in support for existing ones (Table 3.9 on page 140). Although of great benefit to the climate, these have also been criticised for environmental, efficiency or economic reasons.

▼ **Table 3.9** Low carbon emission power generation

▲ **Figure 3.60** Solar farm in Wahlsberg, Hessen, Germany

Solar power

Benefits: Powered by the Sun's light (a renewable source). Can be used on both a large scale or by individuals at a personal level. Technology advancing towards much higher efficiency and for lower levels of sunlight. Cost per watt produced reduced from £50 to around £0.40 between 1975 and 2018.

Disadvantages/criticisms: Concern over large-scale solar farms causing land degradation and habitat loss. Unlike wind farms, there is less opportunity for agricultural use to share the land area occupied. Manufacturing includes emissions of greenhouse gases and the use of hazardous chemicals.

▲ **Figure 3.61** Wind farm, Scotland

Wind power

Benefits: Powered by wind (a renewable resource). Technology is advancing and reducing in price at a speedy rate.

Disadvantages/criticisms: Seen by some as visually unattractive, noisy and disruptive of the natural environment. The wind turbines can become fatal obstacles for birds. Situating and erecting wind turbines and wind farms can destroy delicate soils, environments and habitats. Wind is not consistent and can lead to periods of power-generation inactivity. High winds can result in power generation being shut down to prevent damage to the turbines.

▲ **Figure 3.62** Nesjavellir geothermal power plant, Iceland

Geothermal power

Benefits: Uses naturally occurring heat produced by the Earth (a renewable resource). Worldwide potential is estimated at 2 terawatts. Can be accessed anywhere on Earth (although some areas have better conditions to produce higher amounts of energy). Plants can be built and hidden sub-surface to reduce impact. Recent technological advances have made more resources exploitable while reducing costs.

Disadvantages/criticisms: Initial construction costs are very high and profitable resources for large-scale generation limit locations. Underground greenhouse gases can be accidentally released (emissions are recorded as being higher near these plants). In extreme cases, geothermal power stations can cause earthquakes.

▲ **Figure 3.63** Metz biomass power plant, France

Biomass

Benefits: Created by burning decaying plant or animal waste which is cheap and readily available. If the source is replaced, biomass can be a long-term energy source. Oilseed rape, when treated, can also become an alternative for diesel.

Disadvantages/criticisms: Biomass is only a renewable resource if crops are replanted. When burned, it gives off atmospheric pollutants, including greenhouse gases such as methane. Although the process is cheap, the production of plants and animals for their waste adds considerably to the cost.

▼ **Table 3.9** *(Continued)* Low carbon emission power generation

▲ **Figure 3.64** Hoover Dam HEP plant, USA

Hydroelectric power (HEP)

Benefits: Powered by water (a renewable resource). Long lived. Technology is well understood, reliable and efficient.

Disadvantages/criticisms: Construction involves the emission of large amounts of carbon dioxide (including the manufacture of cement). Construction may cost vast amounts of money that developing nations cannot afford. Valuable agricultural and living space is lost in the creation of a reservoir. Flooding of an area and the imposition of a dam within a river system destroys or disrupts both surface and aquatic environments.

▲ **Figure 3.65** Wave power machine, Orkney, Scotland

Wave power

Benefits: Powered by waves (a renewable resource). No harmful by-products. A consistent source of energy. Technology is advancing and becoming more efficient.

Disadvantages/criticisms: Needs an ocean or sea area to generate power. Machines located on the sea floor disturb the sea floor and alter the environment and habitat for sea creatures. They produce noise which disturbs sea creatures. Surface and sea-floor generators interfere with the ability of ships to pass through the areas where they are located. Traditional fishing areas are disturbed and damage can be sustained, e.g. by catching nets on submerged equipment. Some may disturb the visual nature of the environment.

▲ **Figure 3.66** La Rance tidal power barrier, France

Tidal power

Benefits: Powered by tidal movement (a renewable resource). Tides are more predictable than wind energy and solar power.

Disadvantages/criticisms: Construction involves the emission of large amounts of carbon dioxide (including the manufacture of cement). Construction may cost vast amounts of money that developing nations cannot afford. The imposition of a tidal barrier across a tidal inlet destroys or disrupts both surface and aquatic environments. The large barriers are seen by some as visually unattractive.

▲ **Figure 3.67** Torness nuclear power station, East Lothian, Scotland

Nuclear power

Benefits: This is a well-known technology capable of producing vast amounts of electricity by using nuclear reactions to create steam to power turbines.

Disadvantages/criticisms: Construction involves the emission of large amounts of carbon dioxide (including the manufacture of cement). The cost of construction and decommissioning are high although production of electricity is relatively cheap. The use of nuclear power is a contentious issue. An accident at such a plant could cause wide-scale environmental damage affecting the atmosphere and large areas of land through radioactive pollution. Areas could be uninhabitable for thousands of years.

Geo-engineering

Another method for the control and management of climate change is the use of large-scale geo-engineering techniques. These attempt to influence climate systems by manipulating the environmental processes that govern them. They fall into two types:

- carbon dioxide removal (CDR)
- solar radiation management (SRM).

Carbon dioxide removal

Carbon dioxide removal can be attempted in a number of ways. The best known, and at present most popular, are both methods of carbon sequestration:

- planting large amounts of trees and other plants to create and enhance bio-sinks
- capturing carbon from, for example, industrial processes and then storing it within deep geological formations also makes use of a long-term sink.

Other proposed methods of CDR include the use of artificial trees, enhanced weathering, iron fertilisation and scrubbing towers. Artificial trees are machines created to be similar to natural trees but have enhanced ability to remove carbon dioxide from the atmosphere (predicted potential of 1000 times more efficient at removing carbon dioxide). The carbon is captured in a filter, removed and stored. Enhanced weathering is a chemical approach to geo-engineering. It uses the principles of the natural weathering of calcium and magnesium silicates to transform carbon dioxide into bicarbonate, so removing the gas from the atmosphere. Powdered silicate minerals are spread over land and ocean areas to stimulate and enhance the natural process. In iron fertilisation, iron is introduced (usually in powdered form) to the upper ocean to stimulate the propagation of phytoplankton to enhance their ability to remove carbon dioxide from the atmosphere. Scrubbing towers are structures to remove carbon dioxide from the atmosphere. Air is funnelled into the tower by wind-driven turbines and then sprayed with chemical compounds extracting the carbon dioxide and allowing it to be piped to storage locations, such as geological sinks.

Solar radiation management

Solar radiation management strategies attempt to reflect or divert solar radiation back into space, in reality creating an enhanced albedo for the planet. It is believed that a 0.5 per cent increase in reflective properties could reduce the effect of carbon dioxide doubling by 50 per cent. A variety of methods have been suggested, such as adding aerosols into the atmosphere to increase its reflectivity. It has been suggested that adding 11,800 tonnes of sulphate aerosols into the atmosphere each day would completely offset the additional heat effect of doubling carbon dioxide concentrations in the atmosphere (the equivalent of removing around 200 tonnes of carbon dioxide per day from the atmosphere over a period of 25 years). Due to aerosols being removed from the atmosphere very quickly, if this procedure was relied on solely it would have to be carried out for all time as the build-up of carbon dioxide would still be continuing. Without continued sulphate spraying, the carbon left in the atmosphere would restart its heating with an increased ability due to its increased concentration.

Clouds could have their reflectivity increased by spraying sea water into the atmosphere which thickens the cloud layers and assists in changing the size and distribution of water drops, making them whiter and more reflective. By spraying substances such as silver iodine or dry ice into the atmosphere (cloud seeding), more clouds are formed and this adds to the overall albedo. The release of billions of reflective balloons high into the stratosphere to provide a reflective layer is also proposed. More extreme methods such as setting fire to vast areas of forest so as to release black soot particles seem counter-productive due to the reduction of a natural carbon sink and the introduction of more carbon dioxide into the atmosphere.

Other ideas have been suggested to increase albedo on land such as increasing the use of commercial crops which have a higher reflectivity. A change in the way we construct our towns and cities has also been suggested whereby roofs, pavements, open areas, for example, are planned to have highly reflective surfacing, whether through the addition of paint or through the choice of materials used. One large-scale solution suggested is to cover large areas of land in reflective sheeting, either in deserts or areas where glacial melt has exposed new land. This has a major disadvantage as a large area of land would become off-limits for human use and making vast amounts of plastic sheeting would involve the use of fossil fuels and increase greenhouse gases in the atmosphere.

Some of the most inventive ideas would not take place on Earth but would involve creating lenses, meshes or mirrors in space. The lenses and meshes would allow some solar radiation to proceed towards Earth while filtering out the rest. Mirrors would directly reflect light back into space. Other suggestions include creating a cloud of moondust between the Earth and the Sun, possibly by detonating a nuclear bomb on the Moon's surface to send dust particles into space.

Although some of the geo-engineering ideas are already being put into place, many are seen as impractical at present due either to their cost or unpredictability. The complexity and interconnectivity of the climate system means that geo-engineering has the potential to trigger unforeseen, unpredictable and uncontrollable adverse consequences. Like all climate-forcing mechanisms, geo-engineering may stimulate feedback events that amplify unwanted effects.

Task

1. a) Outline possible methods that might reduce emissions of greenhouse gases.
   b) What are some of the difficulties involved in adopting such strategies?
2. Table 3.9 on page 140 shows methods of low carbon emission power generation. Evaluate the pros and cons of these methods and then rank them in order from most appropriate to least appropriate use. Give reasons for your choices.
3. What is geo-engineering?
4. In geo-engineering, what is the difference between CDR and SRM?
5. Some geo-engineering suggestions, such as artificial trees, scrubbing towers, reflective balloons, covering large areas of land in reflective sheeting, space lenses/mirrors and space dust clouds generated by a nuclear bombing of the Moon, may seem impractical at present due to cost or unpredictability. Evaluate these methods and decide what might make them become practical or necessary.
6. How does recycling help to reduce or slow down global warming?

Managing the effects of climate change being experienced or predicted

There is growing evidence that the world is already experiencing the effects of climate change with extreme weather conditions increasing in their regularity and spatial location. Scotland and the rest of the British Isles are beginning to see a pattern of increased stormy conditions during the winter along with shorter warmer periods during the summer months. In 2015, both Canada and the USA experienced extended periods of sub-zero temperatures and snowfall over extensive areas reaching much further south than would previously have been expected.

In 2018, new records for high temperatures were set around the world and there were prolonged periods of hot temperatures. A series of heatwaves hit North Africa, Asia, North America and Europe. San Diego (on the Californian Pacific coast) experienced record sea surface temperatures of 25.9°C while a cooling current appeared to have shut off. On 11 May 2019 a temperature of 29°C was recorded in North-West Russia near to the Arctic Ocean; a high temperature would normally be around 12°C.

Table 3.7 on page 126 showed predicted worldwide changes that would affect the lifestyles of millions, reduce viable living space and decrease food production. In addition, climate change is already being observed as altering the spatial location of plant types, animals, insects and diseases. Agricultural production, businesses, transportation, energy transfer infrastructure and all of what we see as modern society are under some level of threat.

For developing countries management of, or adaptation to, the changes thrust on them by climate change will be extremely difficult and in some cases impossible. Large-scale changes in climate such as a reduction in rainfall (either in amount or pattern), the raising of temperatures to extremes and sea-level rise may be impossible to manage satisfactorily. In our case study of Tuvalu on page 129 we saw that attempts to protect the country's coastlines would cost billions and would be impossible for the small, developing nation to afford (the estimated cost for a *temporary* sea wall for one *small* atoll could cost over £110 million (approximately $175 million)).

This could lead to starvation, large-scale emigration and abandonment of vast areas of land. Extra pressure would therefore be put on areas (both rural and urban) where resources were available for survival.

Climate change inequalities will exacerbate the differences between affluent and deprived areas within countries and between countries themselves. Stresses could be put on international relationships and within the social framework of individual countries leading to 'resource wars' and civil unrest. International bodies such as the United Nations have suggested that to manage climate change and its consequences, greater responsibility must be taken by all nations to work together, share information and technology, and for the richer ones to take on an additional burden of finance to assist the developing nations and those in greatest difficulty.

Countries need to implement strategies to:

- secure food supply
- find alternative crop types to suit the changing conditions
- react to changes in landscape due to, for example, flooding
- protect valuable and productive land areas (for agriculture, business, housing) or resituate in alternative areas
- adapt to movement of population

- adapt to changing health conditions
- protect national infrastructure including road and rail networks, electricity, gas, water transmission/transportation and sewerage systems
- react to changes in economic circumstances and opportunities.

There will certainly be other issues to be taken into consideration at all levels of management.

In Britain the approach has been bedded on using the best scientific data available and the advice of scientists, business leaders and interested bodies to construct outline plans. These have been prepared to allow for flexibility to deal with levels of uncertainty about what climate change will bring. They promote approaches which allow national, regional, local authorities, businesses and individuals to become pre-prepared, and adjust attitudes and plans to accommodate change.

Although there are individual policies for Scotland, England and Wales and Northern Ireland, there is a generalisation of the approach towards *negative effects* of climate change. Table 3.10 describes actions that have been put in place or proposed.

▼ **Table 3.10** Action against climate change

Climate challenge	Details	Responses
Adapting to flood risks and increasing erosion	Increasing sea levels will place coastal areas under threat of flooding, while increased intensity and number of storms will have the potential to accelerate erosion. Storm surges may damage coastal habitats and agricultural land through their physicality and saltwater intrusion. Increased intensity and amount of rainfall makes flooding more likely, especially on floodplains.	1 Build sea defences such as sea walls to prevent sea water from accessing coastal areas. 2 Maintain or replace sand or pebbles on beaches so as to reduce the speed of incoming waves and lessen their erosive force. 3 Allow local coastal marshes to flood as these can naturally dissipate the force of incoming sea water and reduce damage. In some cases this may mean removing coastal defences that were originally erected to protect these areas. 4 Abandon sea defences for some areas of marginal agricultural land so that it will provide similar defence to response 3 above. 5 Restrict building or development on low-lying coastal areas or where cliffs are of a rock type that will be easily eroded. 6 Improve the effectiveness of natural drainage channels by clearing or dredging rivers and estuaries to allow quick removal of flood waters. 7 Refuse planning permission for buildings on floodplains of rivers. 8 Include planning requirements for climate change-prepared infrastructure including drainage systems (sustainable drainage) for all developments.

▼ **Table 3.10** *(Continued)* Action against climate change

Climate challenge	Details	Responses
		9 Implement planning arrangements that reduce the number of tarred, concreted or pavemented surfaces within built-up areas (including house gardens) to allow water to naturally drain and not overwhelm the artificial drainage system. 10 Increase the level of council tax or other local taxes on houses in areas which are liable to flooding (e.g. already on floodplains) to assist in the creation of flood defences. 11 Higher levels of insurance payments for facilities or housing within areas liable to flood to pay for possible damage while encouraging homebuyers to purchase outside of these areas and remove demand.
Agriculture, food supply, forestry	There has been some decline in agricultural and forest productivity mainly due to the introduction of conditions that allow an increased prevalence of pest, diseases, droughts and the reduction of soil function/fertility.	1 Ensure water supplies through improved maintenance of the supply network and reducing water loss to allow for irrigation of farming land. This is linked to encouragement of responsible water usage by both industrial and domestic users (including the installation of water-preserving technology). 2 Allow for change of crop types which respond better to the changing climate and provide a food source and commercially viable product to preserve the industry. 3 Increase spending on scientific study of pests and diseases to allow for quick response and action. 4 Regular and wide-scale testing and recording of soil fertility to allow for early response. 5 Encourage the planting of a wide variety of crop and tree types to lessen the opportunity for disease and pests to damage the majority of the farmer's or forester's product and thus protect financial and food/resource production.
Health conditions (wellbeing and public health)	Heatwaves are likely to contribute to more deaths in the future especially within the older age groups. Some estimates suggest that by 2040 we will regularly achieve levels of previous heatwaves e.g. 2018, when tens of thousands died prematurely across Europe. Hospitals, homes and care homes as well as other premises may become too hot (many medical facilities are already listed as being so without the impending temperature increases), causing increased ill health especially during summer months. Increased rainfall and temperature may encourage damp and fungus growth stimulating respiratory illnesses.	1 Statutory requirements for new buildings (especially hospitals, homes and care homes) to be designed and equipped to prevent the health risks associated with overheating. 2 Similar requirements to prevent internal temperature and moisture build-ups within buildings to prevent rot and fungus growth. 3 Financial assistance and grants to replace rotted windows, roofs and ceilings and add technology to maintain healthy temperatures. 4 Incentives through lower insurance or tax rates for those who implement the above. 5 Encourage cost-effective passive temperature control methods (usually through design and materials used) rather than air conditioning which can be expensive and add to greenhouse gases through power needed or through their construction.

▼ **Table 3.10** *(Continued)* Action against climate change

Climate challenge	Details	Responses
National infrastructure	Changing climate conditions can increase the rate and level of degradation of the infrastructure, from increasing pothole damage on roads to causing warping on railway tracks, or wind damage to electricity pylons, flooding roads, railways and airports, the disruption of ferry services, and the ability of sewerage systems to deal with increased rainfall and flooding. At present natural hazards account for up to 35% of disruption to electricity supplies, road and rail journeys in the British Isles. Wide-scale degradation of national infrastructure could reduce the living standards of the population and the country's economic potential. Sewerage and flood drains, especially within large cities such as Glasgow, Edinburgh, and London, still rely on systems which are aged and not designed to cope with the increasing levels of rainfall predicted.	1 Systematic approaches by the companies and authorities who maintain the infrastructure to plan and provide for the system to be resilient to climate change. 2 Maintenance of water transfer systems to areas where water shortages are predicted while ensuring that dams and reservoirs are maintained to a high standard to secure storage. 3 Alteration of energy generation to non or low greenhouse gas emitters (Table 3.9 on page 140). 4 Encouraging local/business/facility generation of electricity e.g. solar panels or community wind turbines so as to be less reliant on national distribution. 5 New laws and guidelines put in place to ensure that newly built elements of the infrastructure are protected from the perceived climate changes and conditions, including movement from areas of predicted flooding or coastline incursion. 6 Early replacement and upgrading of sewerage systems and flood drains to counter increased levels of rainfall. 7 Implement planning arrangements that reduce the number of tarred, concreted or pavemented surfaces within built-up areas (including house gardens) to allow water to naturally drain and not overwhelm the artificial drainage system.
Business risks	The location of many business facilities is within the areas predicted to flood (either on river floodplains or at coastal locations). This particularly refers to the larger industrial complexes which need the flatland provided in these locations for easier construction of their facilities. Threats to water supply are a major problem, especially for large-scale manufacturing industries. In the southern half of the UK, major industries already abstract around 46% of the water from catchment areas where the water supply is coming under threat.	1 Businesses are encouraged to plan and implement anti-flood protection at their facilities. 2 Small businesses in particular tend not to have the financial ability to protect themselves effectively or well in advance so incentives are put in place such as grants, low cost loans and reduced insurance costs to encourage flood resilience methods. 3 A mixture of increased taxation and other penalties are placed on industries that fail to replace or improve processes that are wasteful of water (leakages and non-reclamation/recycling). 4 Loans, grants and tax reductions for the implementation of water saving technologies and methodologies. 5 Local control of water to encourage fair shares between industrial and domestic use, backed up with a water pricing system that deters wasteful usage. 6 A response to protect tourism similar to any large-scale industry method mentioned above. Facility owners need to protect their tourism opportunities from flooding through defences, and their water usage in the same ways as any other industry making use of the incentives that make this possible.

▼ **Table 3.10** *(Continued)* Action against climate change

Climate challenge	Details	Responses
	Tourism can be a massive addition to the economy of a local area or a country. In Scotland, for example, its famous links golf courses are under threat from the rising sea level as are many coastal resorts and facilities.	7 With the large number of historical features and sites of natural beauty/scientific interest in places under threat, there is a general international agreement that risk assessment and benefit studies should be undertaken to decide which ones the limited finances should protect, while recording and detailing those which may be lost.
Water supply	In general the British Isles will experience increased amounts of, and more intense, rainfall but there will also be periods of drought conditions especially during the summer.	1 Much of what was stated in the *Business risks* section above applies here (points 3, 4 and 5). 2 Maintenance of water transfer systems to areas where water shortages are predicted while ensuring that dams and reservoirs are maintained to a high standard to secure storage. 3 Reduced water loss through replacement and renewal of pipes. 4 New low-water use technologies to reduce usage (e.g. low-water use toilet flushing systems, automatic shut-off taps, household water reclamation and reuse).

Summary

By now, when answering questions about climate change you should have enough background knowledge to debate and discuss the following:

- physical and human causes
- local and global effects
- management strategies and their limitations.

So, what are the key points to think about? Climate change has been an ongoing process throughout the life of Earth. Before people walked on the planet, the climate system was only controlled and influenced by physical drivers. With the evolution of humans, anthropogenic effects on the climate remained relatively minor until the start of the Industrial Revolution in the mid-eighteenth century. Industrialisation, burning of fossil fuels, the rapid expansion in population, urbanisation and the increased need for resources and space has resulted in human actions impacting on the climate.

Although both physical and anthropogenic drivers continue to influence the atmosphere, it is the human element that has been identified as creating a period of rapid global warming. The changes are happening so quickly that ecosystems and habitats are being altered or destroyed. There are concerns that without action the change in climate may create conditions that, at the least, increase hardships for the human population and, at worst, may take our species towards extinction.

A large variety of management strategies have been agreed to or proposed to help us to either contain the problem or to deal with its resultant issues. Much of the action against climate change has been held back due to disagreements on what methods should be chosen, but recent years have seen a more concerted attempt to find a global strategy.

Some arguments against proposed actions state that they are unfair, difficult to manage or cost too much. It is estimated that if all countries spent 1–2 per cent of their annual GDP on the problem, meaningful change could take place. Unfortunately, there is a growing concern that we may be too late and that whatever we do will not halt large-scale changes to our habitats, and even that the damage already done will lead to uncontrolled temperature rise, environmental shifts and extreme conditions. What could the true cost of our inaction be?

What is certain is that this is no ordinary academic or school subject; it truly affects every one of us. This is a living and developing topic and by the time you read this chapter some of the information given will be out of date and techniques and understandings will have improved. As a geographer, and not just to pass the exam, it will benefit you to investigate and research the topic of climate change further.

Research opportunity

Research and innovation are important parts of the fight against climate change. Some of these have revisited old technology and others show a totally different way of thinking about how we run our lives. Examples are:

- road and pathway surfaces that sense when a car or person is coming and light up or switch on lights that fade after passing (these are being installed in Eindhoven, the Netherlands)

- car park and road surfaces that are also solar panels and act in the same way as above
- ships that are powered by sails, kites and/or solar panel or wind-generated electricity.

You may wish to take some time to research these and other innovations or why not think about things that might have the potential to become part of the fight against global warming. Your ideas may be just what are needed.

Have a balanced but low meat diet (especially less red meat)

Walk, run or cycle short or medium distances and don't take the car

Use public transport and don't use the car for only one person

Buy produce that is locally sourced to cut transport costs

Car share for journeys

Use reusable shopping bags

Fit low-energy light bulbs

Why not work out your carbon footprint and pay for some carbon offsetting?

Don't buy things you don't need (food or other consumer goods)

Don't waste food

Don't just throw out. Reuse and recycle

Don't leave lights on

Choose to vote based on environmental policies

Be aware and keep looking for ways to reduce carbon emissions

Switch off electrical appliances and don't leave them on standby

Recycle old electrical products including games consoles and mobile phones

Become active in encouraging climate protection action

Drive better – accelerate smoothly and slowly, keep to speed limits and maintain steady speeds

Unplug appliances when you are not using them

Don't use drive-through restaurants. Get out of the car and turn the engine off

Don't boil a full kettle when you only need a smaller amount

Buy a low emission or electric car

Use the cool setting on your washing machine

Use reusable containers and buy products with the least packaging

Encourage conference calls rather than long-distance travel

Don't leave mobile devices to charge past their full setting

Take a shorter time in the shower

Don't turn the heating up too far. Wear a jumper and put extra covers on beds

Don't leave doors open

Buy recycled products (even notebooks and toilet paper)

Shower rather than taking a bath

Be climate active

Keep up to date with, and become better informed about, climate change and global warming.

Educate others about climate change and global warming.

Support climate causes that reduce greenhouse gases or fight against global warming.

Encourage, and help improve, energy efficiency.

Join with others or with established groups committed to improving climate conditions.

Encourage your school or college to become climate friendly.

Contact councillors, local authorities, MPs and MSPs and encourage climate friendly approaches and policies.

Be the best you can be and alter your way of life to help save our planet.

▲ **Figure 3.68** What can I do to reduce greenhouse gas emissions?

And remember, this is your life we are talking about. Many of the changes will be happening in your lifetime. Don't just sit back and let others tackle climate change for you, or rely on technology to help out; there are many things that you can do to help solve the problem – take a look at Figure 3.68.

Task

For the final set of questions, it is helpful to break these down into the areas that the SQA states should be studied for climate change. Below is a selection of questions you may experience in an exam. Do not feel that you must answer them all but look for a pattern and see what you need to know to answer them.

Physical and human causes

You may wish to refer to Figures 3.33, 3.35 and 3.36 for the following.

1 Explain, in detail, the physical and human causes of global climate change.
2 Explain the human factors that may lead to climate change.
3 Explain the physical causes of climate change.
4 Explain three types of evidence to show that global climate has changed in the past 10,000 years.
5 Explain how human activities are contributing to the increase in average global air temperature.
6 Global air temperature has increased over the last 250 years. Explain how human activities have affected global warming.
7 There has been an increase in average global temperatures for at least 150 years.
   a) Describe the human factors that may have caused the increase.
   b) Explain how these factors may have led to climate change.

Local and global effects

8 Discuss a range of possible effects of climate change. You should support your answer with specific examples.
9 Discuss the possible impacts of global warming across the world.

10 Explain, in detail, the consequences of increasing temperatures.
11 Explain, with examples, the problems that could be caused by increased sea levels in an area you have studied.
12 Explain the environmental and ecological impacts of global warming in a named area of the world.
13 Explain how global warming can affect food supplies around the world.
14 With regard to Scotland or the British Isles, discuss the possible effects of global warming.
15 Explain how global warming may affect the flora and fauna of Scotland.
16 Do you agree with the following statement? Give detailed reasons for your answer.

   'Global warming will result in increased hunger, thirst, violence and refugees.'

Management strategies and their limitations

17 Explain at least two strategies that have been used to limit the negative effects of global warming.
   a) Explain possible strategies for managing climate change.
   b) Referring to strategies you have studied, comment on their effectiveness.
   c) Outline possible methods to reduce emission of greenhouse gases.
   d) Identify the difficulties involved in adopting these approaches.
18 Evaluate the ways in which it is possible to manage the impact of global warming.
19 Explain how carbon trading operates.
20 Explain the benefits and limitations of a carbon-trading system.

Glossary of key terms

acid rain: rain and any other form of precipitation that has increased levels of acidity. This can have harmful effects on living creatures and plants. Rocks and buildings can also experience erosion due to acid rain. Rivers, lakes and oceans can themselves experience high levels of acidity due to the regular effect of acid rain. The level of acidity within the precipitation is caused by the amount of gases, especially sulphur dioxide and nitrogen dioxide, within the *atmosphere*. The gas level can be influenced by natural occurrences such as lightning strikes and volcanicity. Human activities such as industry, use of motor vehicles and power generation (especially using fossil fuels) have added large amounts of acid rain-creating chemicals to the atmosphere

aerosols: a collection of airborne solid or liquid particles, with a typical size of 0.01–10 µm and residing in the *atmosphere* for at least several hours. Aerosols may be of either natural or anthropogenic origin. They occur naturally, such as in dust from volcanic eruptions or sea spray, and also as a result of human activities, such as those producing smoke or other kinds of air pollution. Aerosols may influence *climate* in two ways: directly through scattering and absorbing radiation and indirectly through acting as condensation nuclei for cloud formation or modifying the optical properties and lifetime of clouds

albedo: the intensity of light (for example, solar energy) reflected from an object, such as a planet or a surface

anthropogenic: caused or produced by humans

anthropogenic greenhouse gas emissions: greenhouse gases, greenhouse gas precursors (chemicals that combine to form greenhouse gases in the atmosphere) and aerosols released by human activity. The main sources of anthropogenic greenhouse gas emissions are burning fossil fuels, clearing forests, agriculture (including livestock), cement manufacture and the use of *chlorofluorocarbons (CFCs)*

aphelion: the point in the path of a celestial body (for example, a planet or a comet) when it is at its greatest distance from the Sun

aquifer: an underground layer of permeable rock, sediment or soil which can contain or transmit groundwater

arthropod: an invertebrate animal having an exoskeleton (external skeleton), a segmented body and jointed appendages

artificial trees: machines created to be similar to natural trees but have enhanced ability to remove carbon dioxide from the *atmosphere* (predicted potential of 1000 times more efficient at removing carbon dioxide). The carbon is captured in a filter, removed and stored

asteroid: small celestial bodies that orbit the Sun. Comprised mainly of rock and metal, they can also contain organic compounds. They are similar to comets but do not have a visible coma (hazy outline or atmosphere) or tail

ataxia: the name given to a group of neurological disorders that affect balance, co-ordination and speech

atmosphere: a layer of gases surrounding a planet or other material body of sufficient mass. Earth's atmosphere is composed of layers of gases surrounding the planet that are held in place by Earth's gravity. The atmosphere is a mixture of (approximately) nitrogen (78 per cent), oxygen (21 per cent) and other gases (1 per cent). Earth's atmosphere thins out with increasing altitude until it gradually reaches space. It is an important part of what makes Earth inhabitable, protecting life on the planet by absorbing ultraviolet solar radiation, warming the surface through heat retention (greenhouse effect), reducing temperature extremes between day and night (the diurnal temperature variation) and through energy redistribution to counteract the difference in *insolation*

attractive targeted sugar-bait traps: these contain active toxic sugar bait (ATSB) which is designed to lure, trap and kill the targeted host vector (the mosquito). The traps are manufactured from easily available ingredients found in tropical and intertropical areas. They contain a scent to lure the targeted vector and a sugar component to encourage feeding. Some traps are designed to hold the vector until it dies; others have insecticide mixed with the bait so that the vector is poisoned and killed

basalt: a common volcanic (igneous) rock formed from the rapid cooling of basaltic *lava* exposed at or very near the surface

biosphere: the part of the Earth and its *atmosphere* in which living organisms exist or where life could be supported. Sometimes referred to as the zone of life

black carbon (BC): fine particulate matter consisting of pure carbon. It is formed through the incomplete combustion of fossil fuels, biofuel and biomass. It can be emitted by both natural and anthropogenic burning and through the soot released. It consists of soot, charcoal and/or possible light-absorbing refractory organic matter

blood cells: human blood contains two types of cell, red and white. Red cells carry oxygen from the lungs to the rest of the body and then return carbon dioxide from the body to the lungs so it can be exhaled. White cells protect the body from infection

blood plasma: the liquid component of blood. It is a mixture of water, sugar, fat, protein and salts. Plasma transports blood cells throughout the body along with nutrients, waste products, antibodies, clotting proteins, chemical messengers such as hormones and proteins that help maintain the body's fluid balance

Brandt Report: written in 1980 by the Independent Commission first chaired by Willy Brandt (the former German Chancellor), to review international development issues. The result of this report provided an understanding of drastic differences in the economic development for both the northern and southern hemispheres of the world

BRICS: in economics, BRICS is a grouping acronym that refers to the countries of Brazil, Russia, India, China and South Africa, which are all recognised to be at a similar stage of newly advanced economic development. This term developed from the acronym BRIC (Brazil, Russia, India and China) to include South Africa when that country achieved the same economic status. In 2013, although the four original BRIC countries covered over a quarter of the world's land area and included more than 40 per cent of the world's population, they accounted for only 27 per cent of the world gross national income

Bti (*Bacillus thuringiensis israelensis*): a naturally occurring soil bacterium that can kill mosquito larvae present in water

calcination: the process of heating a substance to high temperature but below melting or fusing point. This causes a loss of moisture, reduction or oxidation and the decomposition of carbonates and other compounds. Originally this term was used to describe the driving off of carbon dioxide from limestone to obtain lime. Calcination can also be used to extract metals from their metallic ores

cap and trade: a market-based policy tool to assist in the reduction of greenhouse gas emissions. Countries or businesses are given a quota on the amount of emissions allowed but may trade for *carbon credits* or purchase *carbon offsets* to assist with meeting their targets. See also *carbon trading* and *carbon offsets*

carbon credits: a term for any tradable certificate or permit giving the right to emit one tonne of carbon dioxide or an amount of another greenhouse gas that creates the equivalent warming effects of one tonne of carbon dioxide

carbon cycle: a series of processes by which carbon is recycled and reused throughout the Earth's systems. It involves the incorporation of carbon dioxide into living tissue by photosynthesis and its return to the *atmosphere* through respiration, decay of dead organisms and vegetation and fossil fuels. The cycle comprises a sequence of events that assist in making the Earth capable of sustaining life

carbon offsets: a method where a reduction in emission of greenhouse gases is generated (bought or invested in) in order to compensate for or to offset an emission made elsewhere

carbon sequestration: the removal of carbon dioxide from the *atmosphere* by placing it into *carbon sinks*

carbon sinks: natural or artificial reservoirs that store chemical compounds containing carbon. Natural sinks include some rock types, oceans and trees (forests). Oceans absorb carbon dioxide through physicochemical and biological processes. Terrestrial plants, including trees, absorb carbon dioxide through photosynthesis. There are some artificial techniques that capture carbon dioxide and store it (often by storing/trapping it in the Earth's crust). The removal of carbon dioxide from the *atmosphere* by carbon sinks is known as *carbon sequestration*

carbon taxes: levies imposed on fossil fuels aimed at reducing the production of greenhouse gases

carbon trading: a scheme where countries (or businesses) buy and sell carbon permits as part of a programme to reduce carbon emissions. Countries/businesses are given limits on the amount of emissions they are allowed. If they need to go above their limit, they may purchase spare permits (*carbon credits*) from countries/businesses that have produced less than their quota. See also *cap and trade*

catchment: the area of a river and all of its tributaries. It is the area from which a river system obtains its water

chlorine: a chemical element with atomic number 17. It is a yellow-green gas under standard conditions. Chlorine is a strong oxidising agent. A chlorine atom reacts with an *ozone* molecule, taking an oxygen atom with it (forming hypochlorite) and leaving a normal oxygen molecule. It is estimated that one chlorine atom can destroy over 100,000 ozone molecules before it leaves the stratosphere

chlorofluorocarbons (CFCs): any of several organic compounds containing carbon, fluorine and chlorine. Used as aerosol propellants, refrigerants and solvents and in the manufacture of rigid packaging foam. CFCs released into the *atmosphere* rise into the stratosphere, where solar radiation breaks them down; the chlorine released reacts with *ozone*, depleting the ozone layer. Since stratospheric ozone absorbs much of the Sun's ultraviolet radiation, decreased stratospheric ozone levels could lead to increased ground-level ultraviolet radiation, threatening life. International agreements in the 1980s agreed to the phasing out and banning of CFCs

cholera: an infectious and often fatal bacterial disease of the small intestine, causing severe vomiting and diarrhoea, which can lead to extreme dehydration and death. It is contracted by eating food or drinking water contaminated with the bacterium *Vibrio cholerae*

clathrate: a chemical substance consisting of a lattice that traps or contains molecules

climate: the average pattern of the elements of *weather* prevailing in an area over a long period of time (usually around 30 years or more to create meaningful averages). Although more complicated descriptions of climate can be made, it is usual to describe seasonal patterns of precipitation and temperature

climate change: a statistically proven (evidence-based) alteration of the *climate* system over an extended period of time, regardless of what has caused its modification. In addition to this, it may also include a redistribution of the frequency or intensities of *weather* events around the average conditions

climate drivers: factors that influence the shaping and changing of the *climate* system (also known as climate forcings or climate forcing mechanisms)

climate feedback: an interaction between various processes of the *climate* system. Change is stimulated and the result triggers changes in another process that in turn will influence the initial one. Positive feedbacks amplify the effect of *climate change* while negative feedbacks bring about change in the opposite direction

climate proxies: preserved physical characteristics of the past that are used to infer *climate* conditions from their known properties to reconstruct the climatic conditions of times before scientific recording and record keeping

climatologist: a person who studies or practises the study of climate

cloud seeding: a technique used to stimulate or enhance precipitation or cloud formation. To accomplish this, dry ice or silver iodide particles are released into the *atmosphere* or existing clouds

composite indicators: compiled using a number of indicators, to provide a broader, more balanced description of the level of development of a country or region. See also *Physical Quality of Life Index (PQLI)* and *Human Development Index (HDI)*

constructive boundary: where lithospheric plates part allowing magma to rise through the gap in between. New crust is formed when this molten rock reaches the surface and cools. New crust has been constructed

core sample: a cylindrical section of a substance. These are obtained by drilling with special hollow drills that collect a stratified sample of the material from the surface inwards. The sample can then be inspected and analysed

cryosphere: areas of the Earth's surface covered by ice, including ice sheets, glaciers, sea, river and lake ice, frozen land and snow cover

dam: a barrier built across a river or its valley that impounds or holds back rivers to form a lake or reservoir

DDT: a colourless and almost odourless insecticide

Deccan Traps: one of the largest volcanic features on Earth, located on the Deccan Plateau of west-central India. This is a *flood basalt eruption* feature active around 58 to 62 million years ago. At present they cover an area of 500,000 km^2 with a volume of 512,000 km^3, although they are believed to have been around 1.5 million km^2 in size before plate tectonic movement and erosion reduced this

degradation: the deterioration of the environment through depletion of resources such as air, water and soil. The destruction of ecosystems and the extinction of wildlife. It is defined as any change or disturbance to the environment perceived to be undesirable

dendrochronology: the science of dating events, environmental changes and archaeological artefacts by counting and analysing the patterns and sizes of tree rings

destructive boundary: where lithospheric plates collide and one of these plates is forced downwards. As the rock descends the increased heat and friction cause it to melt. This causes destruction of this part of the plate

development: refers to the status of a nation or region's economic situation, society, industry, life style, standard of living, health, education, level of social justice and levels of contentment (quality of life). It can be charted through change over time

development gap: a term used for defining the differences between the most and least advanced countries. It is another way of referring to nations that were defined by the First, Second and Third World statuses. It defines how far apart countries are in terms of *development*, economy and education. The development gap also refers to the hemispheric divide between the North and South

drainage basin: the catchment area of a river and of all its tributaries. It is the area from which a river system obtains its water

endemic disease: a disease that is habitually present in an area due to permanent local causes

endorheic basin: also called an internal drainage system, it is a *drainage basin* or *watershed* that does not flow to one of the Earth's major oceans

enhanced greenhouse effect: the strengthening of the greenhouse effect through human activities (also known as the anthropogenic greenhouse effect)

enhanced weathering: a chemical approach to *geo-engineering*. This technique uses the principles of the natural weathering of calcium and magnesium silicates which transforms carbon dioxide into bicarbonate and removes the gas from the *atmosphere*. Powdered silicate minerals are spread over land and ocean areas to stimulate and enhance the natural process

epidemic: any infectious disease that develops and spreads rapidly to many people. It is usually used to describe an outbreak in an area that is not generally expected to host that disease

evaporation: the process where a liquid is changed into a gas. An example is the oceans being warmed by heat from the Sun and this energy causes water to be turned into vapour which rises into the *atmosphere*

evapotranspiration: the sum of *evaporation* and *transpiration* from the Earth's land and ocean surface and the movement of this water vapour into the *atmosphere*

extinction event: an incident which results in the extinction of one or more species in a relatively short period of time. There is a widespread and rapid decrease in the amount of life on the planet. This is usually as the result of a catastrophic global event, a natural disaster or a sudden change in environmental conditions (also known as an extinction level event, mass extinction event or biotic crisis)

feedback effect: in this case, the consequences of climate forcing which may amplify or diminish its effect. Positive feedback would increase the effect whereas negative feedback would decrease it

First World: a term first used in the 1970s to describe the developed world, including western Europe, North America, Japan and Australia

flood basalt eruptions: large-scale volcanic activity that covers a large area and takes place over an extended period of time. The *lava* can cover hundreds of thousands of kilometres and last thousands or possibly millions of years. This activity can take place on land or in the ocean floor. Due to the large amount of lava, huge volumes of volcanic gases such as sulphur dioxide and carbon dioxide are released. The release of these gases can cause acidic rainfall and drive *climate change*. Flood basalt eruptions such as the *Siberian Traps* (251 million years ago) and the *Deccan Traps* (58 million years ago) have been suggested as the causes of mass *extinction events*

fossil fuels: naturally formed materials that store potential energy that can be released to create power or as heat energy. They are normally formed due to the decomposition and decay of dead organisms that have been buried and contain a high percentage of carbon; examples are coal, oil and gas

fumarole: a vent in a volcanic area, from which smoke and gases arise

gene drive: a *genetic engineering* technology that can spread a particular suite of genes throughout a population or species

genetic engineering: the manipulation of an organism's genes using biotechnology

geo-engineering: the deliberate large-scale manipulation/modification of an environmental process that affects the Earth's *climate* systems, in an attempt to counteract the effects of *global warming*. Geo-engineering techniques fall into two main categories: carbon dioxide removal (CDR) and solar radiation management (SRM)

global dimming: a decrease in the amount of solar radiation reaching the surface of the Earth. This may be as a result of natural processes but it is becoming more regularly used to describe the results of the addition of aerosols to the *atmosphere* by human sources

global heat budget: the usable energy maintained by the Earth as a balance between energy received by the planet (input) through *insolation* and that which is radiated back out into space (output) (also known as the global energy budget)

globalisation: the increasing interconnection of the world's economic, cultural and political systems

global warming: seen by some as being synonymous with *climate change*, it is more accurately described as

the more recent worldwide increase in temperatures believed to be created by human activity that releases *greenhouse gases* into the *atmosphere*. In particular this refers to the period of global industrialisation

global warming potential (GWP): a statement of the amount of heat a greenhouse gas would retain within the *atmosphere* when compared to a similar mass of carbon dioxide

greenhouse effect: the manner by which certain gases warm the *atmosphere* of a planet by capturing or trapping heat that would otherwise have escaped into space; this acts to warm the planet's atmosphere

greenhouse gases: these absorb infrared radiation, trap heat in the *atmosphere* and contribute to the *greenhouse effect*. Carbon dioxide and *chlorofluorocarbons* are examples of greenhouse gases. It is estimated that average global temperatures would be around 32°C lower than at present without the effect of greenhouse gases. Liquid water and life as we know it would not exist under these conditions

gross domestic product (GDP): the total value of all finished goods and services produced by a country in a year, usually expressed in amount per head of population

gross national income (GNI): the total value of all finished goods and services produced by a country in a year plus all net income earned by that country and its population including overseas investments. GNI/capita (US$) is an economic indicator showing the wealth of a country, which in turn shows level of *development*

Gulf Stream: a warm ocean current in the Atlantic originating at the tip of Florida and flowing along the eastern coastline of the USA before crossing the Atlantic Ocean towards western Europe

hard engineering: projects that involve the construction of artificial structures to prevent a river from flooding

host: a human, animal or plant on or in which a parasite lives

Human Development Index (HDI): a composite indicator which uses an adjusted income per capita (income with consideration to the purchasing power), educational attainment (combination of adult literacy rates and average number of years of schooling) and life expectancy at birth. The scale is from 0 (worst) to 1 (best). HDI was devised by the UN to describe human *development* (both economic and social) within and between countries

hydrological cycle: also known as the water cycle, hydrologic cycle or the H_2O cycle, it describes the continuous movement of water on, above and below the surface of the Earth. The mass of water on Earth remains fairly constant over time but the partitioning of the water into the major reservoirs of ice, fresh water, saline water and atmospheric water is variable depending on a wide range of climatic variables

hydrosphere: combined mass of water in all its forms found on, under and above the surface of the planet

ice age: a period of geologic time in which the Earth's *climate* sees a dramatic drop in surface and *atmosphere* temperatures resulting in episodes of extensive glaciation. These episodes of glaciation ('glacials' or 'glacial periods') alternate with periods of relative warmth known as *interglacials*

impact assessment: a cost–benefit analysis that balances the positive and negative features of a development

indicators of development (IoD): used to measure economic, social, demographic, political and cultural statistics for individual countries. IoDs can be used to measure change over time in a country and differences between countries

indoor residual spraying (IRS): a well-established control method for mosquitoes. The insides of houses are sprayed with an insecticide that will kill the vector mosquitoes if they enter a property. Usually all internal walls and ceilings of the building are treated. If carried out correctly this is seen as a highly effective control

infant mortality rate: the number of babies, per 1000 live births, who die before the age of one

inputs: any items or objects entering a system. In nature an example would be moisture entering a system, such as precipitation

insolation: the amount of solar radiation energy received on a surface during a given time. Although taken from the Latin *insolare*, meaning exposure to the Sun, it is often shown as meaning INcoming SOLar radiATION which, although inaccurate, is a good method of understanding its general connotation. Also known as solar irradiation

interglacials: periods of relative warmth lasting several thousands of years during an ice age, separating glacial periods

Intergovernmental Panel on Climate Change (IPCC): the international body for assessing the science related to *climate change*. It was set up in 1988 to provide policymakers with regular assessments of the scientific basis for climate change, its impacts and future risks, and options for adaptation and mitigation

iPhoot doctors: coined to describe a form of primary healthcare strategy where local people are trained in basic medical techniques augmented by mobile phone technology to supply healthcare in places where medical provision is sparse. The name merges the name of popular mobile platforms with the Chinese 'barefoot' doctors on which the basic principles of the strategy are based (also known as smartfoot doctors)

iron fertilisation: a form of *geo-engineering* where iron is introduced (usually in powdered form) to the upper ocean to stimulate the propagation of phytoplankton and enhance their ability to remove carbon dioxide from the *atmosphere*

jet stream: narrow bands of fast-moving air (normally 160–320 km/h but can reach speeds as high as 480 km/h). These assist with the rapid transfer of energy around the world. Jet streams meander around the globe in a west to east direction. They are usually found between 9 and 16 km in altitude, around the height of the tropopause. Jet streams are usually a few hundred kilometres in width with a thickness of around 5 km. They are caused by temperature differences between tropical and polar air masses

K–T Event: the Cretaceous–Tertiary mass *extinction event* (or Cretaceous–Palaeogene extinction event) which occurred around 65 to 66 million years ago. Approximately 70 per cent of all species then living on Earth disappeared within a very short period, including dinosaurs and many species of plants. Often linked to a large meteorite which impacted in the area now known as the Yucatán Peninsula in Mexico although this may not have been the only incident to cause this

lava: molten rock (magma) once it has reached the Earth's surface through a volcano or fissure, vent or crack

Legionellosis: a bacterial disease which may cause pneumonia. It can take the form of a mild respiratory illness or be severe enough to cause death. The bacteria are found in ponds, hot and cold water taps, hot water tanks, air conditioning and cooling towers as well as soil and are spread in the air. Records for this disease go back to 1947 but it takes its name following an outbreak in Philadelphia in 1976 when people attending a convention of the American Legion were affected. It is referred to as Legionnaires' disease. The bacterium is named *Legionella pneumophila* and more recently the official name changed to Legionellosis

least developed countries (LDCs): these are the least developed countries of the world (a sub-set of LEDCs). The LDCs are not only economically deprived and underdeveloped but struggle to improve on their position

less economically developed countries (LEDCs): a broad term defining those countries with a low standard of living and a low HDI when compared with the more developed countries of the world

levee: a natural or man-made embankment at the side of a river. Natural embankments are formed during the process of flooding, whereas man-made embankments are usually built to prevent flooding

life expectancy: the average number of years to which a newborn baby is expected to live, for example 79 in Scotland and 67 in Kenya

lithosphere: the crust and the uppermost mantle which constitute the hard and rigid outer layer of the Earth. It varies in thickness, being on average around 55 km thick beneath the oceans and up to about 200 km thick on continents

Little Ice Age: although not an actual *ice age*, this is the name given to a period from around 1300 to 1870 (*climatologists* contest the starting point and duration) when the northern hemisphere cooled, following a medieval period of increased temperatures. It was at its height between 1600 and 1800. Much colder winters were experienced in Europe and North America

magma: hot fluid or semi-fluid material below or within the Earth's crust from which *lava* and other igneous rock is formed on cooling. Magma also contains crystals, rock fragments and dissolved gases. Magma is between 700 and 1300°C. When magma rises through vents and cracks and appears on the surface of the Earth it is known as lava

magma plume: an upwelling of unusually hot fluid or semi-fluid rock (*magma*) within the Earth's mantle that rises towards the surface. This thin rising column of magma reaches the *lithosphere* and spreads out laterally, stretching and moving the crust above it. Weaknesses form, allowing the magma to rise to the surface. The term 'hotspot' is used to describe the surface activity and location of a magma plume. Magma plumes are thought to be responsible for the formation of the Hawaiian Islands and to have caused *flood basalt eruptions* such as those in the *Siberian* and *Deccan Traps*

malaria: a serious and potentially fatal disease of humans and other animals caused by parasitic protozoans. The disease is transmitted via the bite of the female *Anopheles* mosquito. Mostly found in tropical and subtropical regions

meteor: a meteoroid that burns up as it passes through the Earth's *atmosphere*. Colloquially these are often referred to as 'shooting stars'

meteorite: a meteoroid that passes through the Earth's *atmosphere* and collides with the surface of the planet

Milankovitch cycles: cyclical movements related to the Earth's orbit around the Sun. There are three of them: eccentricity (stretch), axial obliquity (tilt) and precession (wobble)

MINT: an acronym referring to the economies of Mexico, Indonesia, Nigeria and Turkey

mitigation: in this case, a human intervention to reduce the sources or enhance the sinks of *greenhouse gases*

more economically developed countries (MEDCs): the term given to describe the group of countries that are the most economically developed in the world

Mosquirix: see *RTS,S*

mosquito traps: devices designed to attract, then either trap or kill, mosquitoes

newly industrialising countries (NICs): countries in the developing world that have undergone rapid industrialisation since the beginning of the 1970s

North Atlantic Drift: an extension of the *Gulf Stream* (the most important ocean current system in the northern hemisphere), bringing warm waters from the Caribbean north-eastwards. It is responsible for moderating the *climate* of western Europe (also known as the North Atlantic Current or North Atlantic Sea Movement)

North–South Divide: the virtual gulf in *development* that separates the economically developed countries of the North from the less economically developed ones in the South

ocean acidification (OA): the process by which the addition of carbon dioxide to the oceans reduces the overall pH

octenol: a chemical of formula $C_8H_{16}O$, also known as 1-octen-3-ol or mushroom alcohol. It is a chemical that attracts biting insects such as mosquitoes

orogenesis: the process through which mountains are formed following the collision of two tectonic plates. When material is forced upwards, it can cause mountain belts such as the Himalayas or the Alps. Where one plate is less dense than the other, the less dense plate may be forced downwards (subducted) underneath the other, while the more dense one is crumpled to form mountains. The cracks in the crust allow melted material from the subducted plate to make its way to the surface, resulting in volcanic mountain chains. An example of this is the Andes in South America

outputs: any items or objects leaving a system. In nature an example would be all the moisture which leaves a river system, such as by *evaporation* or entering the sea. In manufacturing, outputs could be the product, waste or even profit

ozone: also known as tri-oxygen, it is the triatomic form of oxygen (O_3). It differs from normal oxygen (O_2) in having three atoms in its molecule (O_3). It is a very pale blue gas with a distinctively pungent smell. In the *troposphere* it is created both naturally and by photochemical reactions involving gases resulting from human activities ('smog'). Tropospheric ozone acts as a *greenhouse gas*. In the *stratosphere* it is created by the interaction between solar ultraviolet radiation and molecular oxygen (O_2). Stratospheric ozone plays a decisive role in the stratospheric radiative balance. Its concentration is highest in the *ozone layer*

ozone hole: a phenomenon in which, every year during the southern hemisphere spring, a very strong depletion of the *ozone layer* takes place over the Antarctic region, caused by human-made chlorine and bromine compounds in combination with the specific meteorological conditions of that region

ozone layer: the stratosphere contains a layer in which the concentration of *ozone* is greatest, the so-called ozone layer, which extends from about 12 to 40 km. It protects all life from the Sun's harmful radiation, absorbing 97–99 per cent of ultraviolet light, which otherwise would damage exposed life forms near the surface. Ultraviolet (UV) radiation can lead to varying degrees of skin cancer. This layer is being depleted by human emissions of chlorine and bromine compounds

palaeoclimatologist: a person involved in the study of climatic conditions in the geologic past

palaeoclimatology: the study of climatic conditions, and their causes and effects, in the geologic past

Pangaea: a super-continent that existed during the late Palaeozoic and early Mesozoic eras. The name expresses that this was the 'one land' area in existence at that time. This landmass was located in the southern hemisphere and existed from approximately 300 million years ago until plate tectonic movement caused it to start to break apart around 200 million years ago

Panthalassa: a vast, unitary, global ocean that surrounded the super-continent *Pangaea* during the late Palaeozoic and early Mesozoic eras. Also known as the Panthalassic Ocean

perihelion: the point in the orbit of a celestial body (for example, a planet or comet) when it is closest to the Sun

permafrost: ground (soil or rock and including ice or organic material) that remains at or below 0°C for at least two consecutive years. Most permafrost is located at high latitudes although some may be found at high altitudes in lower latitudes

photosynthesis: a process used by plants (and some other organisms) to create nutrients for continued survival. Using sunlight, these nutrients are synthesised from carbon dioxide and water. A by-product of this process is the generation of oxygen. The process in plants usually involves the green pigment chlorophyll

Physical Quality of Life Index (PQLI): a composite indicator of *development* that combines infant mortality, life expectancy at age one and basic literacy, on a scale from zero to 100

phytoplankton: microscopic algae that form an essential component of the marine food chain. These single-celled plants provide nourishment to many marine species and they also play an important role in regulating the amount of carbon in the *atmosphere*

pollen analysis: used to determine the species present; also known as palynology

positive feedback effect: when a small influence on a process has the effect of causing a larger/additional change in the same direction. It is a process that amplifies the initial disturbance

poverty: a complex term that has different levels of interpretation, but usually means a general scarcity of material possessions or money. Absolute poverty or destitution refers to the deprivation of basic human needs, which commonly includes food, water, sanitation, clothing, shelter, healthcare and education. Relative poverty is defined contextually as economic inequality in the location or society in which people live

proboscis: an elongated appendage from the head of a creature. In insects (such as the mosquito), the term usually refers to tubular mouthparts used for feeding and sucking

pteropod: any of the opisthobranch molluscs comprising two orders (*Thecosomata* and *Gymnosomata*) and having the lateral portions of the foot expanded into wing-like lobes used in swimming. Often referred to as sea snails or sea butterflies

P–Tr Event: the Permian–Triassic extinction event, which occurred around 252 million years ago and estimates suggest that, as a result, around 96 per cent of marine and 70 per cent of terrestrial vertebrates became extinct, as well as a mass extinction of insects. Believed to be the Earth's most extreme *extinction event*

Quaternary Period: a subdivision of geological time running from approximately 2.588 million years ago to the present time. It is divided into two epochs, the Pleistocene and the Holocene. The Pleistocene (2.588 million years ago to 11.7 thousand years ago) was generally a period of wide-scale glaciation and *interglacials* with cyclical growth and retreat of continental ice sheets (periods of planetary cooling and warming). The effects of these *ice ages* resulted in large-scale changes to the landscapes and environments of the planet, with continental glaciers

reaching as far as 40° latitude. The Holocene (11.7 thousand years ago to today) is viewed by many as an interglacial and as being relatively warm within the context of the whole Quaternary Period

radiocarbon dating: a technique used to work out the age of organic materials (for example, wood, bone or cloth) by measuring the amount of the radioisotope carbon-14 (^{14}C) contained within. ^{14}C is acquired from the *atmosphere* (for example, by a living plant through *photosynthesis*) and decays to the nitrogen isotope ^{14}N. The rate of decay of ^{14}C allows the date of the living existence of a sample to be derived (also known as carbon dating or carbon-14 dating)

radiosonde balloon: a battery-powered instrument package raised into the *atmosphere* by a weather balloon. It measures numerous aspects of the atmosphere as it rises, including temperature and pressure, and transmits the information to a ground receiver by radio

rain shadow: an area having very little precipitation due to a barrier, such as a mountain range, blocking the prevailing winds and causing moisture to be removed from the air before reaching the other side of the barrier. When the prevailing winds reach the barrier, they are forced upwards and cooled, causing the moisture within the air to condense and fall as precipitation. The air has now had its moisture removed and is dry. After crossing the barrier, the air is also heated as it descends. As the air warms, it takes in moisture from the land over which it passes. Simply put, air reaching the land on the far side of the barrier brings no moisture and removes the moisture that is there. This causes the area over which the air passes to become very dry: a 'shadow' of dryness is cast behind the barrier (mountain range).

receptor: a protein molecule embedded within the plasma membrane surface of a cell that receives chemical signals from outside the cell

rich industrial countries (RICs): a group of the most developed countries of the world, in some ways similar to the 'North' countries

RTS,S: the malaria vaccine furthest along in development globally. It is the first ever licensed malaria vaccine and the first vaccine licensed for use against a human parasitic disease of any kind. Its trade name is Mosquirix

salivary gland: discharges fluid secretion and saliva

satellite: 1. a natural celestial body orbiting a planet, for example the Moon. 2. a human-constructed device orbiting a planet, moon or other celestial body for investigation/observation, scientific study, military purposes and/or communications

schistosomiasis: a disease contracted by contact with parasitic worms released from infected freshwater snails. Symptoms include diarrhoea, abdominal pain and blood in excrement and urine. Long-term infection may result in liver damage, kidney failure, cancer of the bladder and infertility. In children it may cause stunting of growth and increased chances of learning difficulties (also known as bilharzia, snail fever and Katayama fever)

scrubbing towers: structures proposed for use in *geo-engineering* through which carbon dioxide would be removed (captured) from the *atmosphere*. Air is funnelled into a large confined space within the tower by wind-driven turbines. After spraying with chemical compounds which react with the carbon dioxide in the air, the gas reacts and forms calcium precipitates and water, all of which are then piped to storage locations

Second World: a term referring to the former socialist industrial states (formally the Eastern Bloc), mostly the territory and the influence of the Soviet Union. Following the Second World War, there were 19 communist states and, after the fall of the Soviet Union, only five socialist states remained: China, Cuba, Laos, North Korea and Vietnam

short-wave solar radiation: because of its heat, the Sun gives off radiation with a short wavelength that contains higher amounts of energy. Solar energy enters our *atmosphere* as short-wave radiation in the form of ultraviolet (UV) rays and visible light

Siberian Traps: a large region of igneous rock in Siberia, Russia, formed from a massive volcanic event around 251 to 250 million years ago when the area was located in northern *Pangaea*. Evidence suggests these eruptions were at their strongest for approximately 1 million years but that volcanic activity probably continued in a lesser form for up to 9 million years

snail fever: see *schistosomiasis*

soft engineering: projects that use natural resources and local knowledge of, for example, a river to reduce the risk posed by flooding

speleothems: mineral deposits formed from groundwater in underground caverns. Generally composed of calcium carbonate dissolved from surrounding limestone, for example *stalactites* and *stalagmites*. Stalagmites and stalactites can be used as *climate proxies*

stalactite: a tapering structure hanging from the roof of a cave, often creating a shape similar to an icicle. Formed from calcium salts deposited by dripping water

stalagmite: a mound or tapering column rising from the floor of a cave. Formed from calcium salts being deposited by dripping water

starvation: not having enough to eat

stratosphere: the layer of the *atmosphere* between the *troposphere* and the mesosphere. It is located at an altitude of between approximately 13 km and 50 km

sulphurous acid: a chemical compound that appears to exist only in the gas phase. It is an intermediate state in the formation of acid rain from sulphur dioxide

summer solstice: midsummer, the time of the longest day: this is at around 21 June in the northern hemisphere and 22 December in the southern hemisphere. It is the time when the planet's tilt is most inclined towards the Sun. The Sun is at its northernmost part of the sky in the northern hemisphere and southernmost part in the southern hemisphere. At noon the Sun is at its highest point above the horizon

sunspots: dark, irregular spots on the surface of the Sun (its photosphere). These are temporary areas of concentrated magnetic field. Associated with solar magnetic storms, solar flares and coronal mass ejections. The number of sunspots in a year varies in a relatively predictable way

sustainability: see *sustainable development*

sustainable development: this has been defined in many ways, but the most frequently quoted definition is from *Our Common Future*, also known as the *Brundtland Report*: 'Sustainable development is development that meets the needs of the present without compromising the ability of future generations to meet their own needs'

tectonic plates: the rigid outermost shell of Earth (the crust and upper mantle), which is broken up into slabs called plates. On Earth, there are seven or eight major plates and many minor plates, ranging in size from a few hundred to thousands of kilometres wide. The thinnest plates are usually the youngest created and are found under the oceans, hence the name oceanic plates. The thickest tend to be beneath the continents and are known as continental plates (also known as lithospheric plates or plates). Continental plates mostly comprise granitic rocks made up of relatively lightweight materials. Oceanic crust is much denser and heavier, composed mainly of basaltic rocks

thermal equator: a line which circumscribes the Earth and connects all points of highest mean annual temperature for their longitudes. The parallel of latitude of 10°N, which has the highest mean temperature of any latitude (also known as the heat equator)

thermohaline circulation: a large-scale, density-driven circulation in the ocean, caused by differences in temperature and salinity. 'Thermo' refers to temperature and 'haline' to salt content. In the North Atlantic the thermohaline circulation consists of warm surface water flowing northward and cold deep water flowing southward, resulting in a net poleward transport of heat. The surface water sinks in highly restricted sinking regions located in high latitudes (also known as the global ocean conveyor or great ocean conveyor belt)

Third World: over the last few decades, the term Third World has been used interchangeably with the terms least developed countries, Global South, LEDC and developing countries to describe countries that have struggled to attain steady economic development

Tiger economies: also known as the Asian Tigers or Asian Dragons, it is a term used in reference to the developed economies of Hong Kong, Singapore, South Korea and Taiwan. These nations and areas were notable for maintaining exceptionally high growth rates (in excess of 7 per cent a year) and rapid industrialisation between the early 1960s and 1990s

tipping point: the threshold at which a series of small changes or incidents becomes significant enough to cause a larger, more important, sudden or destructive change. In terms of climate, there are individual

thresholds for different climate elements. For global temperature, the United Nations has stated that a 2°C increase above the pre-industrialisation level would be the point at which climate change would become so rapid (or irreversible), or it would set in train a series of feedback processes, that mitigation attempts would become futile

Total Ozone Mapping Spectrometer (TOMS): an instrument placed on satellites with the intention of measuring ozone values within the atmosphere. It is also able to track sulphur dioxide emissions from volcanic eruptions

transnational corporations (TNCs): businesses that are not restricted to operations within one nation; often referred to as multi-nationals. These companies can develop disproportionate economic and social power

transpiration: the process of water movement through a plant and its evaporation from its leaves

troposphere: the lowest layer of Earth's atmosphere. This layer is the area suited to life as it exists on Earth. It contains the majority of the atmosphere's mass (around 88 per cent) and around 99 per cent of its water vapour

tsunami: a very large ocean wave caused by an underwater earthquake or a volcanic eruption. It increases in height as it reaches shallower waters. Tsunamis can cause great destruction and loss of life in coastal zones

vaccine: a biological preparation that generates active acquired immunity to a disease

vector: an organism, such as a mosquito or tick, which carries disease-causing micro-organisms from one host to another

water cycle: see hydrological cycle

water-related diseases: diseases which occur due to micro-organisms and chemicals in drinking water; they have part of their life cycle in water, have water-related vectors and are carried by aerosols containing certain micro-organisms

watershed: an area or ridge of land that separates waters flowing to different rivers, basins or seas

weather: the state of the atmosphere at a particular location during a short period of time. Weather is made up of a number of elements including temperature, pressure, wind direction, wind speed, humidity, cloud type, cloud cover, visibility and precipitation. On Earth, weather is mostly a phenomenon of the *troposphere*

winter solstice: the time of the shortest day, around 21/22 December in the northern hemisphere and 21 June in the southern hemisphere. The time when the planet tilt is most inclined away from the Sun for the hemisphere in which it is experienced. It is the time of the year when the Sun is at its southernmost point in the sky in the northern hemisphere, or northernmost point in the southern hemisphere. At noon the Sun is at its lowest height above the horizon

World Health Organization (WHO): the directing and co-ordinating authority for health within the United Nations. Responsible for providing leadership on global health matters, shaping the health research agenda, setting norms and standards, articulating evidence-based policy options, providing technical support to countries and monitoring and assessing health trends

Index